ECOLOGY AND CONSERVATION IN PAPUA NEW GUINEA

(A symposium held at the Wau Ecology Institute on 4th November 1975)

Edited by

K.P. LAMB

and

J.L. GRESSITT

WAU ECOLOGY INSTITUTE PAMPHLET No. 2 1976

(Cover by A.D. Johnston)

Ecology and Conservation in Papua New Guinea: A Symposium Held at the Wau Ecology Institute on 4th November 1975 (Wau Ecology Institute Pamphlet, 2)

Edited by K. P. Lamb and J. L. Gressitt

ISBN 978-9980-945-87-7

University of Papua New Guinea Press and Bookshop
PO Box 413
University PO, NCD
Papua New Guinea

CONTENTS

INTRODUCTION

K.P. LAMB[1] and J.L. GRESSITT[2]

The newly independent nation of Papua New Guinea[3] is richly endowed with natural resources of many kinds. Both the terrestrial and marine environments comprise ecosystems of great diversity and abundance of species and individuals of plants and animals. On land the great range of altitudes, the complex topography, the geological diversity and other factors contribute to a wealth of flora and fauna exceeded in few other areas. The marine environment, likewise benefiting from its tropical situation in a setting of great diversity of coasts, reefs and islands, presents a greater productivity and potential than earlier appreciated.

At this early stage in the development of this nation great pressures exist for exploitation of the rich resources - from potential developers on one hand and from the nation and the people for income on the other. Thus it is critical that careful planning be exercised to avoid wasteful exploitation and destruction of non-renewable resources. Because of insufficient knowledge of the general ecology of the region, and even of all the more abundant forms of life, there is great danger of serious upsets in the functions and natural balances. Wise steps toward conservation at this stage may help avoid mistakes made by many other countries before ecology was adequately appreciated. Also it may help to preserve many unique and specialized elements of the biota which may be vulnerable to the effects of modern society.

This symposium represents an attempt to encourage interdisciplinary interation and to foster spread of available information. .....

Education of the population must proceed at the same time as the factors of the environment are studied and elucidated. ...

The symposium was organized by Wau Ecology Institute to celebrate the completion and dedication of its new laboratory, named for Dr. J.J.H. Szent-Ivany. The symposium took place on 4 November 1975, the day after the dedication.

---

[1] Foundation Professor of Biology and Dean of Science, University of Papua New Guinea; Chairman of the Board of Management, Wau Ecology Institute, 1971-75.

[2] Director, Wau Ecology Institute.

[3] Papua New Guinea comprises the eastern half of the island of New Guinea together with the Bismarck Archipelago and Bougainville Island. It formerly consisted of the Australian Territory of Papua (former British New Guinea) plus the Trust Territory of New Guinea (formerly German New Guinea).

The institute was fortunate in being able to assemble many of the government officers and scientific workers in the country who are concerned with some of the above critical problems and factors.

In the meeting, the papers were not read in full, as they had been circulated in advance among the participants. Thus most of the time was spent in discussion. The symposium was followed the next day by a workshop, mainly to discuss the future role of Wau Ecology Institute in the light of urgent national problems.

* * * * * * * * *

ENVIRONMENT, DEVELOPMENT AND PAPUA NEW GUINEA

by

Honorable Stephen Tago, M.P.
Minister for Environment and Conservation

It gives me pleasure to have been invited to contribute in some small way to the discussions which will take place at this seminar.

I am only sorry that the business of the Parliament prevents me from actually being present. I hope that my comments will be of some value and that the conclusions you reach will be of assistance in our combined endeavour of serving the best interests of our people and the nation.

At this point, I would like to also take the opportunity to express our appreciation for the long and distinguished service given by Dr and Mrs Szent-Ivany to our country. Their dedication and deep concern for the future of Papua New Guinea will long stand as an inspiration for our young scientists.

We have generally taken our environment for granted. This makes the task facing me as Minister a Difficult one, for what is generally taken for granted is unlikely to be a policy issue. This is changing, however, as we increasingly realise that there needs to be a deep and lasting commitment of our people to maintaining and also improving our supply of natural resources.

This principle is central to one of our national goals:

"The natural resources and the environment of Papua New Guinea should be conserved and used for the collective benefit of the people; and should be replenished in the interest of future generations.

The natural resources including land, forests, birds, animals, fish, water, sea, air and minerals should be used wisely in the interest of the total development of all citizens throughout Papua New Guinea, and ......
The environment and its sacred, scenic and historical qualities should be conserved and replaced for the benefits of future generations".

The government has defined the framework for our development - our Eight Aims and National Goals. These emphasise that development must be for every man and woman - and not just for the growth of things such as G.N.P. Development must be geared to the satisfaction of our needs beginning with the basic needs of the rural people who make up our majority.

Thus development is a whole: it is a total process involving our values and our culture - it includes the natural environment, social relationships, education, production, consumption and our well-being.

Development for Papua New Guinea will spring from our heart - our success will rely on our strength, our resources and our vision of the land or society we want to create. This is the vision of our constitutional planners and those who assisted them in defining our goals. It is the beginning of the society which will be Papua New Guinea.

In this process of development environmental concerns are a real matter, *our survival*. As I have indicated previously, our real development depends on proper care of the environment. At the same time, our environment and its resource base, respected and used with imagination, will satisfy our needs. It is the supplier of our natural resources - forests, land, water, animals, fish and so on. These two ideas are both a warning and a promise, reminding us that we can exploit nature for our long term benefit *or* to the point where its ability to remake itself and function properly will be threatened.

If we follow the last course, we will all be the losers. The determining factors are two ... on the one hand, the quantity of resources and the laws of nature, ... and on the other, how we exploit these resources.

With these thoughts in mind, I would like to suggest that environmental management and development have one and the same purpose:

. of securing a good standard of living and quality of life

. of maintaining a stable, productive and enjoyable environment and

. of keeping our future options open.

To this end we have defined a number of environmental principles - a charter of environmental rights and obligations - to serve as a national focus. These have yet to be considered by the National Executive Council, but at this point, I would value your comment, criticism and amendments.

1. *All Papua New Guineans have a fundamental right to live in freedom and equality in an environment that provides for our physical and spiritual well-being, and we have a solemn responsibility to protect and improve that environment for our children and for future generations.*

2. *The natural resources of Papua New Guinea including the air, water, plants and wildlife and especially examples of the different kinds of plant and animal communities must be protected for present and future generations by careful planning and management.*

3. *The capacity of the environment to produce vital renewable resources must be maintained and, wherever practicable, restored or improved.*

4. *We have a special responsibility to safeguard and wisely manage the heritage of wildlife and its habitat. Nature conservation including wildlife must therefore receive importance in planning for economic development.*

5. *We must use our minerals and other non-renewable resources wisely, thinking of our future needs, and we must share the benefits coming from the use of these resources among all Papua New Guineans.*

6. *Discharge of heat, poisons or other substances must be prohibited if it damages our plant, animal and human communities beyond repair.*

7. *Papua New Guinea must do everything possible to prevent pollution of the seas by substances that are likely to be dangerous for human health, to harm plants and animals living in or near the seas, or to stop Papua New Guineans from enjoying and using the seas.*

8. *Papua New Guineans must work towards the kind of economic and social development that will provide good living and working conditions now and in the future.*

9. *Papua New Guinea should seek financial, technical and other aid to help with problems caused by natural disasters such as volcanoes and earthquakes and by harmful conditions of underdevelopment such as poor sanitation.*

10. *Stability of prices and adequate earnings for primary commodities and raw materials are essential to environmental management since economic factors as well as the ecological process must be taken into account.*

11. *Papua New Guinea should ensure in its relations with other countries that they contribute to our present and future development rather than removing development opportunities. We should also work towards agreements with other countries to meet the national and international consequences of applying environmental measures.*

12. *Papua New Guinea should seek assistance from more developed countries and from international organisations to include environmental safeguards in our development planning.*

13. *Development planning must consider economic, social and environmental factors together so that the requirements of development do not conflict with the need to protect and improve the human environment for the benefit of all Papua New Guineans.*

14. *Rational planning must be used in order to avoid conflict between the needs for development and the need to protect and improve the environment.*

15. *Planning must be applied to human settlements and urbanisation with a view to avoiding adverse effects on the environment and obtaining maximum social, economic and environmental benefits for all.*

16. *Resettlement schemes and other demographic policies should be used in areas where too many people or too few people make it difficult for development to take place and for environmental quality to be maintained.*

17. *Appropriate national institutions must be entrusted with the task of planning, managing or controlling the environmental resources of Papua New Guinea with the view to improving environmental quality.*

18. *Science and Technology, as part of their contribution to economic and social development must be applied to the identification, avoidance and control of environmental risks and the solution of environmental problems.*

19. *Education in environmental matters for the younger generation as well as for adults is essential so that all Papua New Guineans are able to contribute responsibly to protecting and improving the environment either as individuals or through their communities, businesses or other work. The mass media of communications must be used to help with environmental education.*

20. *Scientific research directed towards solving Papua New Guinea's present and likely future environmental problems must be encouraged. We must support information programmes in order to benefit from up-to-date scientific information and experience from all countries. Environmental technologies should be obtained where possible through foreign assistance.*

21. *Papua New Guinea has the responsibility of ensuring that activities here do not cause damage to the environments of other states or to areas beyond the limits of our jurisdiction.*

22. *Papua New Guinea shall assist in the development of international law regarding the liability and compensation for environmental damage caused by its activities to areas outside its jurisdiction.*

23. *Papua New Guinea must contribute its own system of values to the development of international criteria and standards so that these will be in harmony with our chosen way of life.*

24. *Papua New Guinea should co-operate fully with other nations through multi-lateral and bilateral agreements or other means in order to control the adverse environmental effects of activities in all spheres.*

Development is intimately tied to the formation of economic or monetary value of resources, natural or human. In this process we try to convert materials and energy to more desirable forms and kinds of products. We can do this wisely, or with little concern for the future. If we fail to manage wisely, we will get a spoiling, if not a stopping, of our natural systems and of our communities. It seems to me that such a state of affairs is more often the combined result of a number of factors which originate in:

- Population growth, in terms of size and distribution;
- Urbanization, which unless it is restricted to serve the special vocation of cities, generates an unmanageable burden of services and other problems.
- Industrialisation, which if it is unplanned and uncontrolled generates a whole range of social and economic differences and problems of wastes and pollution;
- Technological change, which if we are not careful about it, will probably exaggerate the inequalities developing in our society. I recognise that there is something very

attractive about, big, clever technology, but we must be careful to judge its costs, including dependency by others on a few, (usually expatriates).

. The assumption that natural resources and water and air in particular, are "free" goods not requiring accountability and responsibility for their continued availability.

If we want to make our future rather than just wait for the fate which will come to us, we have to plan and design our resources development process remembering all these points, otherwise we may be poorer rather than richer at the end of the development process.

For this reason development decisions must be made jointly in terms of social, environmental, economic and technical considerations.

This is shown diagramatically below:

SIMPLIFIED RESOURCES MANAGEMENT

PLANNING PROCESS

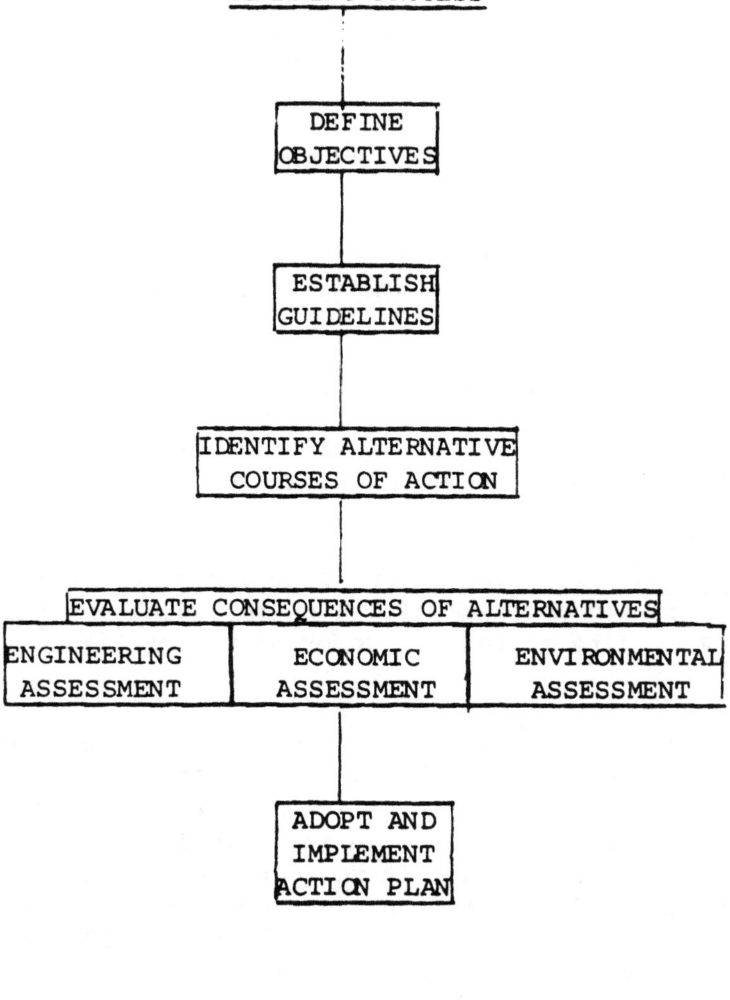

The mechanisms and administrative procedures to do this effectively cannot be put into operation overnight. Procedures suitable for Papua New Guinea have to be worked out, the technical expertise to evaluate environmental aspects of proposed policies, legislation, plans and projects needs to be built up and mechanisms through which the general public can voice their opinions must be established.

Initial steps have been taken - The N.I.D.A. guidelines for feasibility studies on all new projects require the consideration of environmental and social aspects; and the more detailed requirements for forestry and manufacturing activities have already been defined. Others will follow.

In the case of some projects more detailed environmental studies will be required as indicated in the industrial policy statement recently released by the Minister for National Development.

In essence, the objectives of these studies are to ensure that:

A) The effects of the project on the environment are examined;

B) Different ways of accomplishing the project objectives are examined and assessed in order to reduce the effects on the environment;

C) The ways of protecting the environment from bad effects are proposed and evaluated;

D) The project is examined in relation to the present land use and potential regional development of the area;

E) The interested public and people in the area affected are consulted on the environmental effects; and

F) Once the project starts a continual watch is kept in order to warn about problems which might arise.

Important in this approach is the principle that prevention is better than cure - for if any development project does not take into account the effects on the local environment it usually results in future, unforeseen social and economic costs.

The recommended safeguards developed during the environmental study must be taken into account when implementing the project. We have no desire to have our environment contaminated or wantonly abused. We have also been careful to include in our guidelines not only the physical and health aspects, but cultural considerations as well. We are concerned that a development project does not adversely affect our culture, but supports and enhances it. Hence at all stages during the decision making process for a project, the people should be consulted.

The recognition of people's rights is likely to produce a more acceptable pattern of development. This means that the advantages and disadvantages of a particular project must be explained to the community and its wishes respected.

Much more needs to be done. For example, there is considerable technical expertise in the Government service, but scattered through various ministries. There is also much expertise in the universities and industry. Furthermore, environmental concerns and problems fall into the areas of concern of many departments. For example, mine tailings disposal is of concern to Minerals and Energy, Fisheries, Agriculture, Water Resources, Public Health to name a few.

To meet this common concern I plan to establish a National Environmental Council which might include technical specialists from various Government departments, the universities, business and the public. This council would be charged with the task of advising me on technical matters related to environmental management and pollution control, the need for detailed environmental studies, and the like.

The success of the Governments programme of environmental management will succeed or fail on our ability to implement in our day-to-day activity, the thoughts of this paper. I am hopeful that the Environmental Council will be of great value in this effort.

A programme to maintain and improve the environment must have a legal basis. Existing legislature controls, for example, about polluting activities, are largely inadequate to meet our present needs. Broad legislation for environmental protection, establishing the Environmental Council, requiring environmental studies in certain cases and regulating pollution of water, air and land and to maintain the quality of the environment, is in preparation.

In addition, my Ministry has been involved in other legislative initiatives. To give two examples, the New Tourism Act recently passed by the National Parliament and the preparation of a draft bill controlling marine oil pollution and the dumping of trade wastes at sea.

Much more work needs to be done in the legislative area, for many acts and ordinances need review with the intent of incorporating environmental principles in them.

If we want effective land-use and resource management procedures, if we want effective measures to reduce environmental disturbances and pollution control, if we want to set standards, we will need a large amount of factual (scientific if you like) information on our eco-systems, that is the inter-relations between the trees, bushes, animals, water creatures, etc.

For Papua New Guinea this information is very lacking, and so our prediction of likely effects of changes must have some allowance for error or safety margin. Similarly, because of day-to-day work pressure it is often necessary for us to accept (carefully we hope) measures developed in other countries as our starting point so that when more information becomes available on the goodness of copied control programs, some changes can be made if necessary.

This does however highlight the need for an active encouragement of research. I am hopeful that one of the considerations of the working party on science policy will give careful thought to the most effective ways and means of encouraging ecological research.

Information obtained as a result of these research programmes needs to be readily accessible if it is to be useful for planning purposes. The Office of Environment has been examining the ways of establishing a central data bank of information. The first part of this, a coding of information on vertebrate animals has been designed with the assistance of a computer programme.

While still incomplete, I ask your assistance at this seminar in examining the attached pro-formas, making suggestions for improvement and at a later stage by contributing your knowledge and expertise to its effective implementation. Once established it will be a service available to all interested people inside and outside the country.

Up until now I have said little on the subject of environmental conservation and protection. The objectives here are basically the same as those of environmental management. The rational utilization of the natural and human resource potential for increased welfare.

We have a series of unique and complex ecological systems in this country and their conservation is of great importance for the present and the future. Without trying to determine future generations' priorities we collectively have a responsibility to keep the options open and prevent the unnecessary loss of unique species and natural areas.

The identification process for the needed reserves is itself a long and difficult one since it involves the getting together of many specialists who have individual views on potential sites; many of our communities also have serious views on the need to protect their heritage. The National Parks Scientific Advisory Committee has been examining potential sites for national parks and has produced a listing of fifty odd priority areas. A Waigani Seminar (1975) paper by Haaentjens also contained such a listing. However, the process from identification of an area to its dedication as a national park, historic site or wildlife management area is a long and complex one.

Of course there are problems associated with establishing and managing these reserves, not the least being financial and the determination of a land-use priority in competition with other forms of development or use.

Another problem involves customary land tenure, but here our approach in the long term may well have strengths which are important to the success of overall environmental education. Because almost all of our land is under customary tenure it means that we must be careful in site selection in the first place.

Secondly, it means that early consultation with the owners of the area must be started to find out about local land use and development ambitions and carefully explain the concept of national parks, wildlife reserves or management areas. Local committees may be set up and where this has occurred a significant participation has been created right through to the establishment of regulations for control. Once an area is established it is essential that the education be carried through, particularly among the children as they will be the future citizens and leaders.

I do not wish to say any more on this part of the subject, I am sure the participants from National Parks and Wildlife will be able to provide full details of these initiatives.

Let me turn now to education. In the past our informal education was necessarily environmental in its content and outlook. We lived within our environment and if we transgressed its thresholds then by bitter experience we discovered the results. But this has changed.

One of the major tasks of our present education system today is to recreate this awareness. An awareness of our responsibility to our common property - air, land, water, the other natural resources and our cultural heritage. And this awareness must be extended throughout the community, to our children, our elders (perhaps the most knowledgeable already), my fellow politicians and our administrators. Only in this way will we be able to respond to the challenges which face us.

Equally well there is an urgent need to develop an adequate supply of well trained nationals to staff the functions which will give results equal to the problems. I am happy to note that a number of initiatives are being undertaken by the Education Department. For example, in developing an external studies course on environmental matters, by this Institute in preparing booklets, radio-talks and so on, the Forestry College, and the universities where environmental and ecologically directed programmes are being discussed. We must increase these efforts for it is only by an aware public that full measure will be accorded environmental concerns.

There are many other issues that could be discussed, areas where this Ministry is taking initiatives, definition of population policy, a workshop to be held early next year on technologies appropriate for village development, wildlife farming and so on. No doubt many of these will be brought up in the course of your discussions.

I look forward to hearing the outcome.

NATIONAL WILDLIFE PROGRAMME IN PAPUA NEW GUINEA

by

N. Kwapena
Wildlife Branch, D.A.S.F.

SUMMARY

The National Wildlife Programme of the Papua New Guinea Government has been set up to co-operate with landowners and those who use wildlife to ensure that any utilization, whether traditional, economic or subsistence is well planned, allows wild population to be sustained, and minimises damage to habitat. Wildlife and habitat resources are assessed and management projects worked out to allow protection, utilization, mitigation of development's effects and the conservation of all species.

FUNCTION OF WILDLIFE DIVISION IN P.N.G.

The Wildlife Division is the Government instrumentality responsibile for designing and implementing the action under the guidelines of the Government policy.

For efficient and effective wildlife conservation the following are essential:

- the ecological, social and economic factors influencing the continued existence of wildlife must be surveyed;
- effective policies and plans for remedial action adopted by the Government and the people;
- Wildlife Management project put into practice at the national, district and village level; and
- a professional staff, highly trained and sympathetically motivated.

WILDLIFE POLICY

A consistent, long term and developing national policy adopted by the Government through the Minister will control:

- commercial, trading and export of Bird of Paradise, crocodiles, butterflies etc.
- overhunting or collecting of wildlife.
- use of shotguns.
- interference with the people's rights to wildlife by traders, collectors, researchers and visitors.
- the destruction of habitat.
- establishment of Government owned reserves, sanctuaries etc.

WILDLIFE RESEARCH SECTION

This is a section specialized in scientific research investigations on wildlife including flora and fauna and the natural habitats in Papua New Guinea. This section carries out surveys and makes assessments of the biology and ecology of wildlife.

A wildlife ecologist in the Wildlife Research Section is responsible for each primary research project. After assessment of each topic in relation to world literature and the particular departmental requirements, it is the responsibility of the ecologist to draw up a research programme. He must take into consideration objectives, priorities, available finance and man-power, and the people's requirements with respect to wildlife, as well as his primary discipline of wildlife ecology.

WILDLIFE MANAGEMENT SECTION

Conservation and management of wildlife resources are inseperabl It is the duty of this group to investigate, design, implement and administer management programmes based upon the best available informati It is this section which must develop the practical skills needed to maintain wildlife populations.

Regional field stations are located throughout Papua New Guinea at the site of special wildlife problems. These stations which are located at Port Moresby, Lake Murray, Bensbach River, Pagwi, Aird Hills, Cape Hoskings, Bubuletta etc. have demonstration projects to show the people how to conserve wildlife.

The Crocodile Management Group is teaching and helping local people and the Government officers how to farm crocodiles using local materials.

The Wildlife Enforcement Group carries out policing of crocodile and fauna laws. The Enforcement Group is responsible for ensuring that no-one breaks the law in selling protected fauna and oversized crocodile skins.

This group ensures that Birds of Paradise plumes of all species, seven species of birdwing butterflies and the large crocodiles over 20 inches belly-width are not collected, taken, hunter, sold or exported without the permission of the Conservator of Fauna. This group also ensures that non-protected insects and other wildlife dealers for export and import must obtain and produce appropriate permits and licenses.

WILDLIFE MANAGEMENT AREAS AND RESERVES

Throughout Papua New Guinea local Wildlife Committees have been formed. The functions of the Wildlife Committees in each district of Papua New Guinea are:

- to advise the Wildlife Branch and the Minister on their wildlife problems.

- to control and manage the harvesting, collection and hunting of wildlife. Make wildlife laws on the management areas.

- seek assistance from the Government for sustained yield farming of wildlife (e.g. cassowary, wallaby, deer, crocodile, butterfly and orchids).

- to ensure that they receive royalty payments from their wildlife.

In Papua New Guinea Wildlife Management Areas and Reserves have been set up as in the following:

Tonda Wildlife Management Area

Pokili Wildlife Management Area

Baniara Protected Area etc.

Other newly proposed Wildlife Management Areas are:

Garu Wildlife Management Area

Ialibu Wildlife Management Area

Maprik Bird of Paradise Management Area

Dilava-Kubuna Management Area

Wallaby Management Area

In summary the Wildlife Management Section includes the following activities:

- Pilot projects, e.g. crocodile farms and training programme, cassowary farming, birdwing butterly, insect trading, deer and wallaby meat projects.

- Wildlife management areas, wildlife committees, communication teams, etc.

- Law enforcement, smuggling, licenses, permits and so forth.

WILDLIFE TRAINING AND EDUCATION

Wildlife training and education are important. In Wildlife Division, courses are being held for national parks rangers, forestry students, agricultural diplomats, police officers, wildlife officers and the University graduates of Papua New Guinea.

Many people from overseas and inside Papua New Guinea visit wildlife set ups. The Wildlife Education is responsible for making wildlife information leaflets available and taking visitors around.

In summary this section includes staff training in wildlife management, school education programme, literature and adult education aids.

PROVINCIAL WILDLIFE ORGANISATION

At the present time the implementation of the wildlife work at the district and village level is primarily operated by the District Rural Development Officers using diplomates known as Wildlife Officers.

These operate the practical grass-roots level projects by which the people judge the Government programmes on wildlife.

Wildlife stations are set up to provide facilities and other basic needs for the wildlife work. Some examples are:

- Wildlife Research Laboratory, Moitaka, Port Moresby
- Balamuk Wildlife Station, Western District
- Moramora Wildlife Station, West New Britain
- Garu Wildlife Station, Talasea, West New Britain
- Cassowary Management Station, Mendi, Southern Highlands
- Pagwi, Lake Murray, etc.

NEW BRITAIN AREA

The main wildlife problems included in this are are wildfowl egg-grounds, wildlife management areas, wildlife habitat destructions and the forest development. Pokili Wildlife Management Area has been recommended to be declared soon. Wildlife committees have been formed in the Hoskins, Talasea and Matupit area.

Control on Wildlife Management Areas, use of wildlife, enforcement of wildlife laws are required in the area. Development of declaring two more Wildlife Management Areas are required. The Wildlife projects including crocodile farming, and other wildlife management projects are required.

WESTERN DISTRICT

Tonda Wildlife Management Area has been declared. The Balamuk Wildlife Station has developed a crocodile farm and some animals of the area are kept in captivity. Wildlife research teams on deer and other wildlife operate from here. The Lake Murray Crocodile Station is holding 600 crocodiles and this number will have to be increased to 1000.

Development of deer farming, deer meat project and the wildlife management projects including crocodile farming are required. The supply of cassowaries from Morehead are required for a cassowary project in Mendi, Southern Highlands.

EAST SEPIK

Wildlife Committees have been formed at Angoram, Keram and Maprik. A large crocodile farm has been established at Pagwi. This farm is a demonstration farm holding 1000 crocodiles. As a result of this farm, the eight village projects have been known to commence.

Survey work on crocodile farming in Murik Lakes has been carried out recently.

Birds of Paradise Management Area has been considered for Maprik. The butterfly farming and insect trading have been started in Maprik area.

WEST SEPIK (Vanimo)

This area has a variety of wildlife including, crocodiles, Birds of Paradise, cuscus, cassowaries and other unique wildlife species. Control of wildlife collecting from the area is required. Movement of animals across the border needed to be controlled.

SOUTHERN HIGHLANDS (Mendi)

The N.Z. Aid Programme has been developed on cassowary husbandry and traditional farming techniques in the area. The cassowary pens at Was have been established and about 20 animals are kept in captivity.

The Wildlife Management Area has been proposed for the people at Ialibu. The Birds of Paradise Project is considered for this area.

GULF DISTRICT

The area is rich in wildlife especially crocodiles. Purari Hydro-electric Scheme has started already. Wildlife officers are required to develop crocodile farms and survey areas suitable for a Wildlife Management Area. Karimui is proposed. Many crocodile farms have been started already. For example Kikori demonstration farm 150 crocodiles, Ihu 150 crocodiles, Baimuri 150 and other small projects are already being developed by the village people.

NORTHERN DISTRICT

This area is active for butterfly farming and insect trading. There are many local people collecting insects and butterflies for business. Some people have started butterfly farming.

The wallaby meat project has been considered for Tufi area.

The main concern of this area is that there are outsiders collecting and exporting illegally butterflies and insects from this area. Some people are dealing with protected birdwing butterfiles.

The crocodile farming projects have been proposed for the area.

MILNE BAY

Baniara Island has been declared protected for the Agile Wallabies which are decreasing in number. The wallaby farming project has been considered for Goodenough Island.

The crocodile farms have been developed at Bubuleta (demonstration farm of 300 crocodiles), Kuiaro 300 crocodiles, and other small ones have been proposed.

MOROBE

Nothing has been developed here at this moment. Projects to be developed include butterfly farming and insect trading, crocodile farming, control of wildlife in the area, enforcement of wildlife laws on Birds of Paradise and other protected fauna, control of outsiders in dealing commercially with wildlife resources.

WESTERN HIGHLANDS AND THE ENGA DISTRICT

Projects needed to be developed in the area include Birds of Paradise conservation, cassowary projects, enforcement of wildlife laws, control of wildlife collecting in the area and a possibility of establishing a wildlife station.

NEW IRELAND AND BOUGAINVILLE DISTRICTS

The crocodile farming, butterfly farming and the Wildfowl Management Area are considered for these two districts.

The butterfly and insect collecting is very active in these two areas.

These are good areas for crocodile farming activities. The area in New Ireland has been found where the wildfowls are numerous and lay thousands of eggs. This wildfowl egg-ground is found close to a mission station.

The commercial collecting of butterflies and insects needs to be controlled. A Wildlife Management Area is required for the wildfowl habitats.

MANUS

Although the area is physically small, there are important wildlife resources present here. Some of the important wildlife species which need managing are the megapodes, wildlife habitat, dugongs (plentiful near the island waters) turtles and others. An officer is required to develop the above.

It is in this area where the greatest development is urgently required if wildlife conservation is to succeed in Papua New Guinea.

ENVIRONMENTAL MANAGEMENT: ARE TEMPERATE ZONE RESEARCH DATA AND GUIDELINES USEFUL IN THE TROPICS?

by

I. M. Johnstone
Biology Department, U.P.N.G.

A typical temperate zone view of the environment is summarized in the statement "Nature's laws are universal" (1). Today I wish to address myself to this subject. Are nature's laws really universal? I believe that the answer is not as obvious as most people think.

If nature's laws are universal, then it must follow that once the principles or laws operating in one ecosystem have been elucidated they will be found to be just as valid in any other ecosystem. Furthermore, management guidelines drawn from these laws should be equally valid everywhere. In practical terms this would seem to imply that, like physicists and the atom bomb, biologists should "do a Manhattan" on some convenient ecosystem. The laws discovered to operate in that ecosystem, then, would enable us to answer all our environmental management questions irrespective of which system they apply to. To a certain extent this has been the approach in recent years (IBP); however, most ecologists are still involved in small-scale problem solving research on some particular ecosystem, this approach seemingly contradicting the belief that nature's laws are universal. I will return to this subject in more detail later.

When dealing with any environmental problems, whether it be resource or pollution oriented, an environmental policy that is based on fundamental objectives is needed so that planners can proceed to a cost-benefit analysis of the situation. A typical set of objectives promulgated by a temperate zone theoretist are as follows (2).

I. Economic

- National economic efficiency
- Regional economic development
- Income redistribution

II. Environmental

- Minimum disruption of natural environment

III. Social

- Political equity
- Prestige
- Acceptability
- Well-being of people

Are such objectives applicable in the tropical zone? I suggest that they are only if

1. The country has a "Western-style" culture
2. The country is economically developed.

For example, in a developing country such as Papua New Guinea, where there is perhaps 90% cash-economy unemployment, one could easily imagine that the basic economic goal of a project would be none of the above alternatives but rather job creation.

Thus, I believe, there is a real danger in accepting, without modification, general cultural goals from other regions of the globe. This danger is illustrated by some recent attempts to quantify the goals of environmental policy. These techniques used allow the status of the environment or the potential impact on the environment to be expressed in terms of a single "Environmental Quality Index" (3). However the process of collecting the masses of raw data required to calculate the index leaves little opportunity to question whether the correct parameters are being taken into account in terms of the real needs of the society or country involved.

For a particular environmental problem, a legal requirement may be passed and technical and ecological alternatives may be available. Two examples are given below (4).

Legislative Issue	Technological Approach	Ecological Approach
Water-quality goal	Use to which water body is to be put by man	Maintenance of natural physical, biological, and chemical integrity of the water
Mode of treatment of pollutants	Primary treatment followed by discharge into environment with assumed assimilative capacity	Recycling and elimination of any discharge into natural waters

Such alternatives may not be adequate, however, to cope with the situation in a tropical country or a non-Western society. For example, in China the approach to sewage is not one of a pollutant but that of a raw material (5) so that western guidelines for sewage discharge quality become meaningless. In a developing country where few people have a reticulated water supply, the major goal, perhaps, should be the delivery of water to more people rather than concentrating limited resources on attaining a high standard of water purity.

A scarcity of highly trained manpower can lead to a particularly insidious problem ... the failure to interpret ecotechnological jargon. A particularly interesting example of this is the concept of "assimilative capacity". The idea of the assimilative capacity of an ecosystem is usually central to the technologist's human-use approach to the environment. By "assimilative capacity" does the technologist mean

. the ability of the system into which the pollutant is injected to transfer that pollutant elsewhere, or

. the ability of the system to dilute the pollutant within itself without the negative feedback loops in the system becoming positive.

If the first interpretation is true as has been claimed (6), then we can only congratulate human nature on being consistent (out of sight out of mind); while on the other hand if the second is correct (7) then we are left with the peculiar fact that the environment is not using all of its resources (i.e. O_2 content).

In the Madang Urban Study (8), for example, it seems clear that the authors view the "assimilative capacity" of Binnen Harbour in terms of its "flushing capacity". They state (p.344) "Because of (the) low tidal movements the pollution assimilative capacity of the harbour will be less than many harbours of similar size". A similar view is also held by the authors of the impact statement on the J.A.N.T. Madang Chip Mill (9). However they go even further, stating that "it would be appropriate to establish the total assimilative capacity of Binnen Harbour and apportion to the company a proportion of this capacity after considering the requirements of the rest of the business, industrial and residential community". What is it they wish to apportion? Clearly the "excess" of oxygen in the harbour is the basic resource under debate, which falls under the second concept of "assimilative capacity". However, the "flushing ability" of the harbour would also be a portion of the resource. It would seem that the concept of "assimilative capacity" is usually thought of in terms of "excess" oxygen and/or "flushing ability". Thus, we are back to square one; since "flushing" is just sending the problem somewhere else and the idea of "excess" oxygen in an aquatic ecosystem is without experimental proof.

Meanwhile, an ecological study (10) of Binnen Harbour has shown that marked environmental deterioration is already evident in the Harbour due in large part to the dumping of household sewage directly into the water. In addition, rather than there being a "flushing" action with the adjacent ocean, winds and currents actually bring floating debris into the harbour from along the coast.

Clearly, then, for an untrained person it would be very difficult to see through the jargon of a term such as "assimilative capacity" and realize that a nonexistent resource may be being apportioned to companies who hope to use it to get rid of their wastes.

Although it is often difficult to transfer developed country guidelines to developing countries I believe there are certain basic "international" ecological laws which can be used to form guidelines. Until recently ecology was largely a descriptive science and could not be used easily to predict the results of alternative management techniques. However, today this situation has changed (11) and the problem of choosing the correct management techniques has largely been reduced to one of correctly defining one's objectives and then selecting the appropriate Principle or combination of Principles which have been recently elucidated (11). For example the basic objective of North American sports fishing is to maximize the harvest from the population of some particular piscivorous fish species (12). To do this one needs to obey Principle 4 i.e. for all categories of resources, when resource availability is already high, the effect of a unit increase in resource availability often decreases with increasing availability up to some maximum. Beyond this maximum there is no further beneficial effect with a further increase in resource. This means one must harvest a particular age class only. To get as much of this one type of fish as possible one would also just have this particular piscivorous species present. On the other hand, the basic objective in a developing country is to maximize the harvest of fish protein, the type of fish being irrelevant. To do this one must take into account two additional principles: Principle 2 which states that no energy conversion system is ever completely efficient (in other words, farm herbivores); Principle 9, which states that the diversity of any community is proportional to the biomass divided by the productivity (in other words, farm a mixture of herbivorous fish).

The problem with employing basic ecological laws is that trained manpower is often not available to convert them to a practical approach to a particular problem. It is more usual to take research data and convert these to a management guideline. This, however, can cause problems since most research data is available only for temperate ecosystems. To give an example, if we wished to determine the size of a conservation area for butterflies we could look at the data of Erlich et al (13) and quite legitimately decide that an area of a few acres would be suitable. However, to use this data for the tropics would be to court disaster. For example, the tropical butterfly *Heliconius ethilla* like many tropical bees, traplines in order to obtain pollen that provides the necessary amino acids. Thus, an area of at least one square kilometer of forest is required to support a population of this species (14, 15). Does this example mean that we should not proceed with any action concerning a tropical ecosystem until we have raw data concerning that system?

Another example illustrates the problem further. For many years it has been readily accepted that tropical and temperate soils are different. In fact it was even believed by many that there existed a unique tropical soil-forming process that led to the formation of laterite. Latosols and lateritic soils erroneously became known as soils in the process of developing into laterite and even relatively recent publications in widely read journals conclude that when most tropical areas are cleared of vegetation they will become worthless brick pavement in a few years (16). In fact, tropical soils can be quantitatively defined only as those soils which lack a significant summer-to-winter temperature variation (17), otherwise soil management techniques tend to be identical to those of similar soils in the temperate zone. On the other hand, rates of mineral cycling and the location of nutrient pools are quite different in many tropical ecosystems compared with similar temperate zone ones (18, 19, 20).

Since a part of an ecosystem, the soil, can be essentially the same in the tropics and the temperate zone, are there whole ecosystems which are similar enough to be treated as identical? Surprisingly enough, the answer is a definite "yes". For example, tropical sandy beaches have faunal and floral assemblages very similar to those of subtropical and temperate beaches, even having the same genera (21, 22, 23). On the other hand, there are ecosystems in the tropics for which there are no even remotely similar parallels in the temperate zone, for example, the mangrove swamps.

I feel that the worst problem in the exchange of information between the temperate zone scientist and tropical decision makers is the quickly-drawn evaluation of some aspect of tropical biology offered by the "one-day stop-over scientist". The conclusions of these visitors must be treated with the utmost caution. An example from my own field, phycology, will clearly illustrate this problem. There is today in phycological circles a concept I call the Britton Hypothesis. In 1945-46 M. E. Britton, during his military service, was able to visit Milne Bay in Papua New Guinea and various other places in the West Pacific. Based on these stop overs, he stated that the algal flora "are rather monotonous ... rich in neither species nor individuals" (24). Such a mistaken conclusion can only be understood by knowing that tropical fresh water habitats can show marked seasonal fluctuations in their algal populations (25) and that the best algal "homes" are not often found next to airports.

In conclusion, I believe that there are a number of basic ecological laws and so support the contention that "nature's laws are universal". However, these principles are not very useful unless a group of highly trained scientists are available to understand and skilfully apply them. Since this is usually not the case in the tropics, guidelines must be formulated essentially on the spot from data collected preferably in the area to be exploited.

Furthermore, because the basic objectives of a developing country and/or tropical country in terms of social wellbeing may not be the same as those of a highly technological temperate zone country, guidelines from these countries must be treated with caution.

REFERENCES

(1) Davis, J. Minister of the Environment for Canada In: Environment Canada (1972) Canada and the Human Environment. Information Canada, Ottawa. 92 pp.

(2) Westman, W. E. (1972) Some basic issues in water control legislation. American Scientist, 60: 767-773

(3) Inhaber, H. (1974) Environmental Quality: Outline for a National Index for Canada. Science, 186: 798-805.

(4) Berry, B.J.L., Horton, F.E. (1974) Urban Environmental Management. Prentice-Hall. 425 pp.

(5) Wortman, S. (1975) Agriculture in China. Sci. Am., 232(6): 13-21.

(6) U.S.D.I. (1970) The National Estuarine Pollution Study. U.S. Senate Doc. 91-58, Washington D.C. 633 pp.

(7) Carter, L.J. (1970) Galveston Bay: Test Case of an Estuary in Crisis. Science, 167: 1102-1108

(8) PNG Department Lands Surveys and Mines (1972) Madang Urban Study. Russel D. Taylor and Partners Pty. Ltd., Boroko. 396 pp.

(9) Beca Carter Hollings and Ferner, Professional Opinion. Environment Protection Madang Timber Project. May, 1972. Report to PNG Government.

(10) Unpublished results of UPNG student field trip to Madang.

(11) Watt, K.E.F. (1973) Principles of Environmental Science. McGraw-Hill, 319 pp.

(12) Bennett, G.W. (1971) Management of Lakes and Ponds. Van Nostrand Reinhold. 375 pp.

(13) Ehrlich, P.R. et al (1975) Checkerspot butterflies: A historical perspective. Science, 188: 221-228.

(14) Ehrlich, P.R., Gilbert, L.E. (1973) Population Structure and Cynamics of the tropical butterfly *Heliconius ethilla*. Biotropica, 5: 69-82

(15) Gilbert, L.E. (1972) Pollen feeding and reproductive biology of *Heliconius* butterflies. Proc. Nat. Acad. Sci. USA, 69: 1403-1407.

(16) McNeil, M. (1964) Lateritic Soils. Sci. Am. 211(5): 96-102.

(17) Sanchez, P.A., Buol, S.W. (1975) Soils of the tropics and the world food crisis. Science, 188: 598-603

(18) Greenland, D.J., Kowal, J.M.L. (1960) Nutrient content of the moist tropical forest of Ghana. Plant and Soil, 12: 154-174.

(19) Nye, P.H. (1961) Organic matter and nutrient cycles under moist tropical forest. Plant and Soil, 13:333-346.

(20) Rodin, L.E. and Bazilevich (1967) Production and mineral cycling in terrestrial vegetation. Oliver and Boyd, 288 pp.

(21) Gonzales-Liboy, J.H. (1971) The zonation and distribution of marine beach macroinvertebrates at Plaza Mani, Mayaguez, Puerto Rico. Unpub. M.S. thesis U.P.R. Mayaguez, P.R.

(22) Wade, B. (1967) Studies on the biology of the West Indian beach clam, *Donax denticulatus* L. I. Ecology. Bull. Mar. Sci., 17:

(23) Pearse, A.S. et al (1942) Ecology of sand beaches at Beaufort, N.C. Ecol Monogr., 12: 136-190.

(24) Britton, M.E. (1948) New Species of *Oedogonium* from Leyte, the Philippine Islands. Am. J. Bot., 35: 715-719.

(25) Unpublished observations of the author.

THE PURARI FISHERIES STUDY - AN ENVIRONMENTAL IMPACT STUDY IN PAPUA NEW GUINEA

by

A. K. Haines
Fisheries Research Station, D.A.S.F., Kanudi

I. INTRODUCTION

This paper is intended to outline the reasons for the Purari fisheries study, and the approach used in the study. It is not designed to give the detailed results which are coming out of the work. Papers on these are in the initial stages of preparation and will not be completed for some time. The brief section on results is only to illustrate the trends which are appearing.

Papua New Guinea is a developing country and scientific work must be related to improving the economy of the country or the quality of life for "the man in the village". For this reason considerable attention is given in this paper to fisheries as an economic resource and fisheries development.

II. REASONS FOR THE STUDY

1. Background to the Purari Scheme

For many years observers have been impressed with the apparent potential of the Purari River for producing hydro-electric power. The Japanese firm of consulting engineers Nippon Koei Co. Ltd. (N.K.), initiated with Papua New Guinea Government approval and conditions, investigations into the hydro-electric possibilities between August 1971 and March 1973 (Snowy Mountains Engineering Corporation (SMEC) - N.K., 1974). Following this, the Papua New Guinea Government, in September 1973, commissioned a study by U.N.D.P. consultants on comprehensive development of the Purari River Basin (Barr, 1974). This report was produced in May 1974 and became known as the "Barr Report". The Barr Report included recommendations for inland and offshore fishery developments and the possible establishment of fish-based industries. It also suggested fishery developments in the dams constructed for hydro-electric purposes. Australian and Papua New Guinean Government talks in August 1973 resulted in a S.M.E.C. review of the N.K. report and subsequent S.M.E.C. and Australian involvement in the scheme (S.M.E.C., 1974).

Early this year the P.N.G., Australian and Japanese Governments agreed that a full technical feasibility study by S.M.E.C.-N.K. would proceed. It is envisaged that this study will take three years. It will examine technical aspects of hydro-electric power production at Wabo and other sites, the transmission of this power to a port site on the coast, and the development of an industrial complex at the port site.

As early as November 1972 cabinet directed that "an ecological study of the effect of the dam and the overall hydro-electric project be carried out at an early date, and certainly before any commitments are entered into to build any such works" (Dyer, 1975). After the fisheries study had commenced, a U.S. consulting firm, Ecological Research Associates, produced a report recommending impact studies on all the environmental aspects of developing the Purari River. Among the recommendations were several outlining detailed fish studies in the proposed inundation zone, the river below, the delta and the Gulf of Papua (Goldman, Hoffman and Allison, 1975).

2. Purpose of Fisheries Work

The Purari fisheries study began in August, 1974. From a resource viewpoint, it is necessary to gain information on the fishery resources of the area and the possible modes of development for these resources. Should the Purari Scheme proceed, the development of fishery resources can be tied in with the overall economic development of the region. The supply of agricultural products and fish to the industrial centre, for example, could be a means for economic benefits to flow to the villagers who wish to retain their traditional life-styles. If, for any reason, it is decided not to proceed with the hydro-scheme, then alternative developments for the area will probably be considered, especially as the expectations of the people have been raised by talk of the Purari Scheme. As the Gulf District is very rich in fish resources, fishery development would play a large part in any alternative proposals, hence it is necessary to acquire accurate information on the fish and potential fisheries of the area.

From an environmental viewpoint, information on the fish ecology is necessary to predict likely effects of developments both on the fishery resources and on the total environment. A number of commercial fisheries are associated with the Gulf of Papua e.g. barramundi (c. 150 tonnes p.a.), prawns (700 tonnes p.a), bluefin tuna (25 tonnes p.a.) and crayfish (100 tonnes p.a.). Any likely changes in, for example, mangrove nursery areas, nutrient flow into the Gulf, or industrial pollution as a result of the scheme must be considered for their effect on existing fisheries. Likely losses of either commercial or subsistence fisheries must be included in the overall cost of the scheme. Conversely, any likely gains can be considered among the benefits of the scheme.

Finally, information on the pre-impoundment fish ecology is important for purely scientific reasons.

A number of other studies have begun, or are about to begin, on the Purari environment, including wildlife, forestry, botanical and soil surveys, and a study on nutrient origins and transport. International aid agencies are being approached for assistance for studies or sediment and water quality, hydrobiolgoy and aquatic invertebrates, particularly crustaceans and molluscs.

3. Purari Study Compared with Studies on Other Tropical Dam Projects

There have been fishery and environmental studies on large tropical dam projects in a number of places. There is considerable biological information available from studies on Lake Brokopondo in Surinam, Volta Lake in Ghana, Lake Kainji in Nigeria, Lake Kariba in Zambia and Rhodesia, the Tungabhadra Reservoir and other schemes in India, and the Mekong Scheme in South East Asia. Other schemes about which the author has as yet no information are Lac de Kossou in Ivory Coast, Lake Nasser in Egypt and the Ord River scheme in tropical Australia.

The information available on the various studies is very valuable, however, the Purari study differs in a number of ways. It is not possible to give the exact dimensions of the proposed Purari developments, however, Petr (1975) has summarized the figures available from the various reports - S.M.E.C., 1970; Barr, 1974; Lang, 1974; S.M.E.C.-N.K., 1974. It seems the dam at Wabo will be comparatively small (approximately 260 Km^2) and with a drawdown of 25 m. In addition, it will be situated in a remote, rugged and thinly populated part of the Gulf District. For these reasons the main fisheries interest is not in the man-made lake as in other schemes. The main concern is with the delta where there is a larger population with some established fishing tradition. The delta supports a vastly under-exploited resource of fish, prawns, and crabs and includes extensive mangrove forests which almost certainly provide nursery areas for the Gulf of Papua prawns and are probably a major source of nutrients for the Gulf waters.

This study also differs from other studies in that the fish of the Purari River are predominantly marine and estuarine forms living in freshwater. In the Asian, African and South American cases the fluvio-fauna has been dominated by purely freshwater fish.

Finally, the emphasis in this study is on feasibility and pre-impoundment, whereas previous studies mostly began after the dam was closed, or a few weeks before, and their emphasis was on the post-impoundment stage.

III. APPROACH USED IN STUDY

1. Ecological

This programme is not a study testing theoretical ecological principles, it is designed to give the basic information necessary for the environmental and developmental purposes outlined above. The precise aims are to:

(a) identify the species present

(b) establish the distribution patterns of the species

(c) determine their food and feeding relationships, and the niches of the various species within the wider ecosystem

(d) establish some idea of relative abundances

(e) gain some understanding of the annual cycle, including migrations

(f) gain information on reproduction i.e. times, places, movements and special requirements

(g) attempt to determine the important environmental requirements of the various species

Ideally, this information should be sought for all species present, in practice, of course, it is necessary to concentrate on those forms of particular ecological or economic significance. Although the aims presented suggest some sort of chronological order, the actual study is a process of gathering data on all aims at the same time e.g. it would be impractical to wait until all taxonomic problems were sorted out before seeking other data.

A variety of gill and seine nets are used for sampling, while some specimens are purchased from villagers. Attempts to use *Derris* root poison have so far failed because the water is either too deep, extensive or fast flowing. The sampling provides specimens for (a) while the catch data give information on (b), (d) and (e). The results must be interpreted cautiously, however, as gill nets sample selectively both for size and species. Active surface-swimming forms such as archerfish (*Toxotes)* for example, tend to be over-represented, while sluggish, bottom-living forms such as gudgeons (F. Elotridae) can be greatly under-represented. The following data are recorded for all fish obtained: identity, length, weight, stomach contents, gonad development and presence of young (in mouth-breeding forms). These data give the basic information for (c), (e) and (f), and supplementary information for the others.

Collecting localities are classified as river (between Wabo and the effluence of the delta), freshwater section of delta, estuarine section of delta, and coastal delta. Within these broad zones as many types of environment as possible are sampled, e.g. coastal sand-bars, mangrove lagoons, main and side channels, isolated and partially isolated billabongs, tributary streams, springs, etc. This approach gives useful information on (b)-(g) e.g. a particular mud-eating cat-fish, *Hexanematichtys carinatus,* is extremely abundant in muddy, semi-isolated billabongs along the river and in the freshwater section of the delta. In the main channel it is present but comparatively uncommon. If only the main channel was sampled it would give a very misleading idea as to the ecology of the species and its importance in the aquatic community. The main approach to the problem of (g) used in this study is this comparative one. The adjacent river systems of the Vailala, Era and Kikori-Seribi Rivers are included within the study for this reason. To give an example of applying this approach for (g), the waters of the Purari are very turbid with suspended sediment. The Kikori River is similar in this respect and has a similar fish fauna - parts of the adjoining Seribi River, however, have quite clear waters. In the Purari and Kikori Rivers the fork-tail catfish (F. Ariidae/Tachysuridae) are dominant, both in biomass and numbers of species. In the clear waters of the Seribi catfish are virtually absent. A reasonable hypothesis is that muddy water is an ecological requirement for the Family. Future sampling may either disprove or corroborate this hypothesis.

2. Developmental

A part of this study is based on village fishing surveys. From an environmental aspect, it is important to know which species are important in the subsistence economy, and exactly how important they are. Resource-wise, it is considered that, to have any chance of success, fisheries development must be consistent with existing fishing traditions, at least initially. The existing commercial fishery exports only 150-200 kgs of barramundi fillets per month from Baimuru, plus unknown quantities of fish and crabs are sold in the local markets. The village surveys will eventually give figures for the existing subsistence fishery. Matters such as who catches the fish; how, where and when they are caught; who consumes them and how much is consumed; what quantities of what species are utilized, and how the fish is cooked and preserved are all important in establishing a picture of traditional village fishing. Should the Wabo dam be built this information will also be useful in settling possible compensation claims.

Work is being done at Kanudi, Daru and in the Sepik on methods of fish preservation suitable for the village situation e.g. salting, drying and smoking. The outcome of this work will be very important to fishery development in the Purari region. It is possible these techniques will eventually reduce dependence on the "freezer system" in fishery development at the village level. A freezer needs sophisticated expertise to build and maintain, it needs regular supplies of fuel, and must be regularly full and with a rapid turnover of frozen fish to be economic. Freezers thus tend to promote increased reliance on complex technology which needs imported expertise, equipment and fuel.

Of the two freezers in Baimuru, one no longer operates because it was uneconomic with the small quantity of fish passing through it. The other, which exports 150-200 kgs of barramundi fillets per month only operates because it is the freezer for the hotel, and fish represent a minor sideline.

The greatest single resource in the Purari which could be developed by "appropriate technology" methods is catfish (F. Ariidae). Catch figures consistently show that for every kg. live weight of barramundi taken, 5-10 kg. live weight of catfish are caught. Experiments on catfish processing are planned in the future.

IV. EMERGING PATTERNS

1. Distribution

About 100 species from 48 Families have so far been recorded in the study. Most are restricted in distribution to one or several of the zones described above. *Hexanematichtys leptaspis* (goldspot catfish), *Lates calcarifer* (barramundi), *Scutengraulis scratchleyi* (freshwater anchovy), *Toxotes chatareus* (archerfish) and *Curcharhinus leucas* (river whaler shark) are the only species which appear ubiquitous The fauna is dominated, both in biomass and number of species (approximately 20), by the Family Ariidae (Tachysuridae), the fork-tail catfish.

An interesting point is the depauperate faunas of the Seribi and Vailala Rivers compared with the Purari, Era and Kikori. The case of the Seribi River has been described above. The reason for the poor fauna (and catch results) in the Vailala is still obscure - future work on the physico-chemistry of the water may clarify matters.

2. Food Chain and Nutrient Cycle

The rivers are fast-flowing, subject to sudden floods, very turbid and with highly unstable bottoms. Perhaps for this reason there is a lack of aquatic vegetation, even in usually isolated billabongs. Chemical data given by Petr (1975) show that the isolated billabongs and side branches do, however, contain water with higher concentrations of ions and probably nutrients.

The complete food chain cannot, of course, be determined until studies have begun on nutrient flow and aquatic invertebrates. The fish can, nevertheless, be divided into their feeding relationships. The consumers of terrestrial vegetation e.g. figs and pandanus fruit fallen into the water *(Hexanematichtys acrocephalus)* and terrestrial animals e.g. insects *(Toxotes chatareus)* depend directly on the terrestrial community. The mud-eaters (numerous species including *Scatophagus argus, Hexanematichthys carinatus, Hemipimelodus crassilabris,* and *Nedystoma dayi*) presumably get a large proportion of their nutrients from broken down terrestrial matter. These particular groups, then, transfer part of the energy fixed by terrestrial primary production into the aquatic section of the ecosystem.

There are also predators of aquatic animals at various levels in the food chain. Most feed largely upon prawns (e.g. *Kurtus gulliveri, Johnius belengeri, Scutengraulis scratchleyi, Cochlefelis spatula* and several other catfish whose taxonomy is still uncertain). Prawns represent a major part of the food chain needing further investigation. Some species feed mainly on other fish (e.g. *Lates calcarifer, Carcharhinus leucas, Stenocaulus kreffti* and several catfish.) Several species are omnivores (e.g. *Hexanematichthys leptaspis*), while one, *Cinetodus froggatti,* appears to feed exclusively on molluscs.

Most of the various categories of consumers can be found at all localities from the river, through the various zones of the delta, to the coast. The actual species vary with the differing zones. Moreover, with the apparent exception of *Cinetodus frogatti,* no species is exclusive in feeding habits.

3. Village Fishing

So far only one village survey has been completed. The village, Akoma, is in the coastal zone of the Purari delta. Three points are particularly noteworthy:

(i) fishing is very much an activity of women and girls.

(ii) the main protein source of the village is crabs, prawns and small fish, particularly gudgeons (Eleotridae), gobies (Gobiidae), perchlets (Chandidae) and small catfish (Ariidae).

(iii) the main fishing methods were with hand traps and lines - gill nets were not used.

V. COMMON ASPECTS OF ENVIRONMENTAL IMPACT STUDIES IN PAPUA NEW GUINEA

People in Papua New Guinea live much closer to nature than those in industrialized countries. Consequently they are much more directly affected by any change in their natural environment. Papua New Guinea's largely unpolluted, unruined environment is its greatest resource - the majority of people can derive all or most of the necessities of life from it without entering the cash economy. Destruction or damage to the environment in this country means a very direct and immediate economic loss to the people concerned. All environmental impact studies carried out in Papua New Guinea would have several aspects in common.

(a) Aim

Any major development must cause changes in the environment, some of which will be adverse. The aim of any impact study should be to provide the decision-makers with the best information possible on likely environmental changes, both good and bad. It is not the role of the study to condemn or support any development - the probable gains and losses should be presented to the decision-makers and it is their job to assess the relative importance of each.

(b) Working Problems

Logistical problems. due to difficulties of communication and uncertainty of transport, must loom large in any study. Data gathering can never be as complete or as regular as one would like, in fact, it is often a case of going into the field whenever possible and gathering whatever data one can.

(c) Scientific Setting

The lack of scientific knowledge and newness of record-keeping mean that little is known of most areas. Most scientific work must begin from scratch e.g. taxonomic problems are a large part of any biological study here.

(d) Economic Interest

Because this is a developing country with limited resources, research should be economically biased. This can mean research directed either towards improving cash income at national or village level, or towards improving or maintaining the quality of life of villagers outside the cash economy.

VI. REFERENCES

Barr, R. L. 1974 Comprehensive River Basin Development (Purari) Project No. PAP/73/010/A/01/31 U.N.D.P. Papua New Guinea Tippetts-Abbett-McCarthy-Stratton Engineers & Architects, New York.

Dyer, K. W. 1975 Environmental Studies for the Purari River Office of Minerals and Energy, Purari Co-ordinating Section File 55-5-2 memo to Purari Officials Committee.

Goldman, C. R., B. W. Hoffman & A. Allison 1975 Environmental Studies Design Purari River Development Papua New Guinea Ecological Research Associates, Davis, California.

Lang, T. A. 1974 Purari River Development and Wabo Project Leeds, Hill & Jewett, Inc. San Francisco.

Petr, T. 1975 The Purari River - Wabo Scheme Comments on the Hydrobiology and Fisheries Development.

Snowy Mountains Engineering Corporation 1970 Assessment of runoff and hydroelectric potential of Papua New Guinea.

Snowy Mountains Engineering Corporation 1974 Review of Nippon Koei Preliminary Report.

Snowy Mountains Engineering Corporation-Nippon Koei Co. Ltd. 1974 Proposal for Consulting Services Feasibility Study for Wabo Power Project. Vol. 1 Technical Proposal Cooma and Tokyo.

COMMUNITY INVOLVEMENT IN CONSERVATION AND ECOLOGY PROGRAMS

D.H. Dale
School of Natural Resources,
University of Technology, Lae.

Abstract: Papua New Guinea is doubly fortunate in that it is richly endowed with natural resources yet not beset by the problems of development which many industrialised countries, particularly those with large populations, must face.

There is no room for complacency however since it is clear that similar problems will afflict this country in due course if there is no forward planning.

Given that substantial funds cannot be diverted into conservation and ecology programs, particularly at this time, it is fortunate that education can be used to provide the vehicle for effective change. Such is the experience overseas. In Australia, for instance, recycling of paper and glass has risen, in less that ten years, from low levels to 50% and 40% respectively. This has come about because Australians have become more concerned about the environment and are now being educated to be concerned for it, particularly in school.

The supreme example of course is provided by China where total involvement of the community has led to the virtual elimination of pestilences such as flies, starlings and bilharzia and where close monitoring of the environment is achieved through the wide-spread practice of "Folk Geology".

What steps can be taken in Papua New Guinea in order to achieve significant and effective community involvement in conservation and ecology programs ?

Three such steps are :-

(i) Creation of a community awareness

(ii) Education and training

(iii) Initiation of national projects and action plans.

Creation of a community awareness

Existing and potential problems and desirable goals need to be clearly identified. The initial lead for this must come from government and particularly from Departments such as Education, Information and Extension Services and Agriculture, Stock and Fisheries.

Greater use of the news media will help create such an awareness. However there seems little doubt that information (e.g. wildlife protection) can be most effectively spread through the community by the children.

Education and training

The education system reaches right across the nation and most sections of the community are associated with some form of schooling. The attention of these schoolchildren should be directed towards conservation and ecology programs and their energies harnessed through activities planned and implemented on a national basis. There is plenty of scope for action in this direction (e.g. science and environmental newsletters, competitions, projects) and valuable environmental and ecological base line data (e.g. concerning vertebrates) could be collected and assembled on a regional basis.

Initiation of national projects and action plans.

Rural and urban life-styles are markedly different in Papua New Guinea. Conservation and ecological problems in these areas are also fairly distinct.

Fortunately both sectors are structured in such a way that the council is the important link between individual and government. Wildlife committees have been formed by many village councils and there is no reason why a wide variety of committees and groups should not be set up in both the rural and urban sectors of the community, with particular support being drawn from the universities, technical institutions, high schools and charitable organisations.

ECOLOGY AND CONSERVATION: FLORA OF PAPUA NEW GUINEA

ABSTRACT

K. J. White,
Department of Forests, Papua New Guinea.

Taking note that the study of the ecology of the Papua New Guinea flora is the essential basis for any serious consideration of effort in conservation, we should ask what has been achieved to date; and what are the directions we should highlight.

On the ecology front a number of broad scale classifications of the vegetation have been made. These are interpretable with classifications in other tropical situations and they provide a preliminary basis for further in depth studies - both of the classification themselves; and of the plant associations which comprise them.

A few examples of this broad nature are Lane Poole (1925), Womersley and McAdam (1957), the CSIRO Regional Resource Surveys, the Archbold Expedition reports, New Horizons (1973), Fords' Atlas of Resources (1974) etc.

Detailed studies are much fewer in number and are generally limited to comparatively small areas and varying degrees of scientific value. Background information is contained in a large number of Department of Forests files and internal reports, and specialist publications as, Percival and Womersley Floristics and Ecology of the Mangrove Vegetation of Papua New Guinea; Ashe's paper on Nothofagus, Walker's work on pollen in highland lake deposits; Gillison on Doma Peaks, and other occasional papers.

As an aid to students and research workers I have attempted a bibliography with the assistance of the Division of Botany and this is attached. I am sure this is not complete but it will form the basis for a wider and more searching literature review.

This then is our state of knowledge of the flora environment, and the basis on which its management must rest. It is not adequate, and the review illustrates the large deficit that must be overcome to place conservation policies and practices on a sound ecological basis; without this their scientific value is sharply diminished. In the meantime the needs of conservation must of necessity be satisfied within:

- the protection provided by natural isolation
- environment management plans in development projects and
- ad hoc reservations based on constrained (or little) scientific background.

In any involvement in conservation there is the element of management practices - these may be natural phenomena or of intervention by man. I have stated elsewhere that I consider there are few if any climax plant communities in the lowlands of Papua New Guinea and this could well be extended to the forest communities of the Highlands. I have further noted the planned developments of forestry, agriculture, power, roading and mining in the forest zone and its logical long term outcome of radically altering the forest system. There is also the wide-scale and evident impact of man in shifting cultivation, hunting and foraging for food and other values on the forest. It is often advanced that there is an wholesome stable relation between Papua New Guinea village communities and the forest. I strongly challenge this. As I see it, man here is not associated with the forest in some beneficial symbiotic liaison but, that the association is a parasitic one. Man dwells on the forest fringe, mining the forest for the small timbers needed for his housing, fuel etc., later converting this denuded forest to horticultural use (Pandanus, Gnetum, Artocarpus) and then to garden crops and, ultimately, the resiliant grasslands. There are of course probing uses of the interior of the forest in hunting, gathering of medicines, foraging etc. The combined impact is to roll up the forest through this creeping fringe occupation and to replace it with the durable and easily maintained grasslands. The very development of these grasslands with their low food yields and their difficulty of management for agriculture with hand tools impels the forest destruction process onwards.

It is interesting to note the reversal of this traditional destructive use in the Eastern Highlands since the intervention of European settlement in the last thirty years. The introduction of iron implements (ploughs, spades etc) has allowed development of agriculture in the grasslands; cessation of tribal fighting has promoted village expansion in the valley floor; cash cropping, and the establishment of round wood forests for fuel, building material etc has almost eliminated the use of the receeding forest fringe. These new forests are largely the result of introduction of _Casuarina oligodon_ into a fallow cropping system. Shrubby regeneration is now covering the hills and if fire is kept out, the effect would be a return to a many species tall closed forest system in 200-300 years.

Clearly the pursuit of conservation falls into a number of activities. These include:

- the development of embracing reserves, which is a normal recommendation of meetings such as this,
- the inclusion of environment management in development projects. As a forester I feel competent to recommend that in forest operations:

  land use planning is an essential component of the project with a highlighting of environ environmental management plans for affected village communities and for flora and fauna purposes as well as for wood production covering:

  - adequate reservation of forested areas which are integrated in an overall plan for village community traditional purposes; watershed protection; fauna and flora reserves; stream protection; cultural and customary use reservations.

  - the restriction of clear felling to those areas which will be subsequently required for follow up use in agriculture and forest cropping and grazing, and for infrastructural needs for roading, towns etc.

  - the retention of the remainder of the forest in a physically structured frame that will minimise adverse impacts.

- pollution control over all aspects of the industry.

- the inclusion of environment management training in all levels of education, primary, vocational, secondary and tertiary.

I personally consider the last, is the most important contribution that can be made in Papua New Guinea.

BIBLIOGRAPHY OF SELECTED PAPERS ON PLANT ECOLOGY IN PAPUA NEW GUINEA

Aland, F.P. (1975). Impacts of agriculture in the lowland rainforest area of Papua New Guinea. Paper presented to Symposium on Man and the Biosphere project. Port Moresby, April 1975, 5 pp.

Archbold, R. and A.L. Rand (1935). Results of the Archbold Expeditions. No. 7. Summary of the 1933-1934 Papuan Expeditions. *Bull. Amer. Mus. Nat. Hist., 68*.

Ash, J.E. (1975). The ecology of *Nothofagus* and mixed species forests on S.E. Mt. Giluwe, Papua New Guinea. *Unpubl. report*, Res. Sch. of Pacific Studies, Aust. Nat. Univ., Canberra.

Brass, L.J. (1941).- The 1938-39 Expedition to the Snow Mountains, Netherlands New Guinea. *Jour. Arnold Arb.*, *22*: 271-342.

Brass, L.J. (1956). Results of the Archbold Expeditions. No. 75. Summary of the fourth Archbold Expedition to New Guinea (1953). *Bull. Amer. Mus. Nat. Hist., 111*: 81-152.

Brass, L.J. (1959). Results of the Archbold Expeditions. No. 79. Summary of the fifth Archbold Expedition to New Guinea (1956 1957). *Bull. Amer. Mus. Nat. Hist., 118*: 1 - 69.

Brass, L.J. (1964). Results of the Archbold Expeditions. No. 86. Summary of the sixth Archbold Expedition to New Guinea (1959). *Bull. Amer. Mus. Nat. Hist., 127*: 147 - 215.

Cartledge, E.G., D.E. Shaw, and D.J. Stamps (1975). Studies in relation to dead patches of *Nothofagus* in Papua New Guinea. *Dep. Agric. Stk. Fish., P. Moresby, Res. Bull. 13*: 1-26.

Coode, M.J.E., and P.F. Stevens (1972).- Notes on the flora of two Papuan mountains. *Proc. PNG Sci. Soc.*, *23*: 18 - 25.

Flenley, J.R. (1967). The present and former vegetation of the Wabag region of New Guinea. *Unpubl. Ph.D. thesis*, Aust. Nat. Univ., Canberra.

Flenley, J.R. (1969).- The vegetation of the Wabag region, New Guinea Highlands. A numerical study. *J. Ecol.*, *57*(2): 465 - 490.

Ford, E. (ed.) (1974).- *Papua New Guinea resource atlas*. Brisbane Jacaranda Press. 56 pp.

Gillison, A.N. (1969).- Plant succession in an irregularly fired grassland area - Doma Peaks region, Papua. *J. Ecol.*, *57*: 415 - 429.

Gillison, A.N. (1970). - Structure and floristics of a montane grassland/forest transition, Doma Peaks region, Papua. *Blumea*, *18*: 71 - 86.

Gray, B. (1973). - Distribution of *Araucaria* in Papua New Guinea. *Res. Bull. No. 1, Dept. Forests, PNG*.

Havel, J.J. (1971). - The *Araucaria* forests of New Guinea and their regenerative capacity. *J. Ecol.*, *59*: 203 - 214.

Heyligers, P.C. (1965). - Vegetation and ecology of the Port Moresby Kairuku area. *C.S.I.R.O. (Aust.) Land Res. Ser. No. 14*: 146 - 173.

Heyligers, P.C. (1966). - Observations of *Themeda australis* *Eucalyptus* savannah in Papua. *Pac. Sci.*, *20*: 477 - 489.

Heyligers, P.C. (1967). - Vegetation and ecology of Bougainville and Buka Islands. C.S.I.R.O. (Aust.) Land Res. Ser. No. 20: 121 - 145.

Heyligers, P.C. (1972). - Analysis of the plant geography of the semi-deciduous scrub and forest and the Eucalypt savannah near Port Moresby. Pac. Sci., 26: 229 - 241.

Heyligers, P.C. (1972). - Vegetation and ecology of the Aitape Ambunti area. C.S.I.R.O. (Aust.) Land Res. Ser. No. 30: 73 - 99.

Heyligers, P.C. (1972). - Descriptions and ecology of the vegetation types of the Vanimo area. C.S.I.R.O. (Aust.). Land Res. Ser. No. 31: 86 - 98.

Heyligers, P.C. (1973). - Vegetation of the low-lands (Papua/New Guinea). Aust. Nat. Hist., 17(12): 404 - 409.

Hope, G.S. (1973). - The vegetation history of Mt. Wilhelm. Unpubl. Ph.D. thesis, Aust. Nat. Univ., Canberra. 460 pp.

Hynes, R.A. (1970). - Nothofagus forest in New Guinea, an ecological introduction. Outline summary progress paper - Botany section 42nd ANZAAS Congress.

Johns, R.J. (1972). - Vegetation. In P.A. Ryan (ed.) 'Encyclopaedia of Papua and New Guinea'. Melbourne Univ. Press.: 1163 - 1170.

Kalkman, C. (1963). Description of the vegetation types in the Star Mountains region, West New Guinea. Nova Guinea, Botany No. 15.

Kalkman, C., and W. Vink (1970). - Botanical exploration in the Doma Peaks Region, New Guinea. *Blumea*, *18*(1): 87 - 135.

Lane-Poole, C.E. (1925). - *The forest resources of the territories of Papua and New Guinea*. Government Printer, Melbourne. 209 pp.

Paijmans, K. (1967). - Vegetation of the Safia-Pongani area. *C.S.I.R.O., (Aust.) Land Res. Ser. No. 17*: 142 - 167.

Paijmans, K. (1969). - Vegetation and ecology of the Kerema-Vailala area. *C.S.I.R.O. (Aust.) Land Res. Ser. No. 23*: 95 - 116.

Paijmans, K. (1970). - An analysis of four tropical rain forest sites in New Guinea. *J. Ecol.*, *58*: 77 - 101.

Paijmans, K. (1971). - Vegetation, forest resources, and ecology of the Morehead Kiunga area. *C.S.I.R.O. (Aust.) Land Res. Ser. No. 29*: 88 - 113.

Paijmans, K. (1973). - Vegetation of Eastern Papua. *C.S.I.R.O. (Aust.) Land Res. Ser. No. 32*: 89 - 125.

Paijmans, K., and E. Loffler (1972). - High altitude forests and grasslands of Mt. Albert Edward, New Guinea. *J. Trop. Geog.*, *34*: 58 - 64.

P.N.G. Department of Forests (1973). - *New Horizons: Forestry in Papua New Guinea*. Brisbane Jacaranda Press, 70 pp.

P.N.G. Department of Forests. - 90/Series files. Forest resource surveys.

P.N.G. Department of Forests. - 125/Series files (Division of Botany): Botanical expeditions and reports.

P.N.G. Department of Forests. - 126/1 Series files (Division of Botany): Ecological studies (special areas).

Percival, M., and J.S. Womersley (1975). - Floristics and ecology of the mangrove vegetation of Papua New Guinea. PNG Dept. Forests, Division of Botany. Botany Bull. No. 8. 96 pp.

Powell, J.M. (1970). - The impact of man on the vegetation of the Mt. Hagen region, New Guinea. Unpubl. Ph.D. thesis, Aust. Nat. Univ., Canberrra.

Rand, A.L., and L.J. Brass (1940). - Results of the Archbold Expeditions. No. 29. Summary of the 1936-37 New Guinea Expedition. Bull. Amer. Mus. Nat. History. 77: 341 - 380.

Robbins, R.G. (1960). - The anthropogenic grasslands of Papua and New Guinea. Symp. Impact Man Humid Trop. Vegn. Goroka (T.P.N.G.): 313 - 329.

Robbins, R.G. (1961). - Montane formations in the central highlands of New Guinea. Proc. Symp. Humid Tropics Vegetation, Tjiawai (Indonesia) 1958. U.N.E.S.C.O.

Robbins, R.G. (1961). - The vegetation of New Guinea. Australian Territories, 1: 1 - 12.

Robbins, R.G. (1961). - The montane vegetation of New Guinea. Tuatara 8: 121 - 134.

Robbins, R.G. (1968). - Vegetation of the Wewak. Lower Sepik area. C.S.I.R.O. (Aust.) Land Res. Ser. No. 22: 109 - 124.

Robbins, R.G. (1970). - Vegetation of the Goroka Mt. Hagen area. C.S.I.R.O. (Aust.) Land Res. Ser. No. 27: 104 - 118.

Robbins, R.G. (1974). - Vegetation. In Ford, E. (1974), loc. cit., pp. 12 - 13.

Robbins, R.G., and Pullen, R. (1965). - Vegetation of the Wabag Tari area. C.S.I.R.O. (Aust.) Land Res. Ser. No. 15: 100 - 115.

Royen, P. van (1963). - The vegetation of the island of New Guinea. TPNG Dept. of Forests, Division of Botany, Lae. 70 pp.

Royen, P. van (1967). - Some observations on the alpine vegetation of Mt. Biota (Papua). Acta Bot. Neerl., 15: 530 - 534.

Saunders, J.C. (1957). - Vegetation. C.S.I.R.O. (Aust.) Div. Land Res. and Regional Survey, Div. report 57/2. Lands of the Gogol Upper Ramu area, New Guinea.

Schodde, R. (1973). - General problems of fauna conservation in relation to the conservation of vegetation in New Guinea. In Costin, R.B., and R.H. Groves 'Nature conservation in the Pacific', Aust. Nat. Univ. Press: pp. 123-144.

Smith, J.M.B. (1974). - Origins and ecology of the non forest flora of Mt. Wilhelm, New Guinea. Unpubl. Ph.D. thesis; Australian Nat. Univ., Canberra. 270 pp.

Specht, R.L., E.M. Row, and V.H. Boughton (Ed's) (1974). - Conservation of major plant communities in Australia and Papua New Guinea. Aust. J. Bot. Supplementary Series, Suppl. No. 7. (July, 1974).

Taylor, B.W. (1957). - Plant succession on recent volcanoes in Papua. J. Ecol., 45: 233 - 243.

Taylor, B.W. (1964). - Vegetation of the Wanigela-Cape Vogel area. C.S.I.R.O. Land Res. Ser. No. 12: 69 - 83.

Taylor, B.W. (1964). - Vegetation of the Buna-Kokoda area. C.S.I.S.R.O. Land Res. Ser. No. 10: 89 - 98.

Wade, L.K. (1968). - The alpine and subalpine vegetation of Mt. Wilhelm, New Guinea. Unpubl. Ph.D. thesis, Aust. Nat. Univ., Canberra.

Wade, L.K. and D.N. McVean (1969). Mt. Wilhelm studies. I. The alpine and subalpine vegetation. A.N.U. Press Publ. BG/1. 225 pp.

Walker, D. (1966). - Vegetation of the Lake Ipea region, New Guinea highlands. 1. Forests, grassland and garden'. J. Ecol., 54: 503 - 533.

Walker, D. (1968). - A reconnaissance of the non-arboreal vegetation of the Pindaunde catchment, Mount Wilhelm, New Guinea. J. Ecol., 56: 445 - 466.

Walker, D. (1970). - The changing vegetation of the montane tropics. Search, 1 (5): 217 - 221.

Walker, D. (1973. - Highlands vegetation (Papua/New Guinea). Nat. Hist., 17 (12): 410 - 414.

Wardle, P. (1969). - Report on a visit to New Guinea. D.S.I.R., Botany Division, N.Z. (report unpubl.).

Wardle, P. (1971). - An explanation for alpine timberline. N.Z. J. Bot., 9 (3): 371 - 402.

White, K.J. (1971). - Lowland rain forest in Papua New Guinea. Paper to Pacific Sci. Assoc. Conf., Bogor. 7 pp.

White, K.J. (1974). - Rain forest: management of the ecosystem: ecological and environment considerations. P.N.G. Dept. Forests, Port Moresby. (Paper to U.N.E.S.C.O. seminar, Bogor, Dec. 1974).

White, K.J. (1974). - Forestry. In Ford, E. (1974). loc. cit., pp. 14 - 15.

White, K.J. (1975). - Planned decelopments in lowland rainforest in Papua New Guinea with special reference to infrastructure, mining and forestry. PNG Dept. Forests Port Moresby. 13 pp. (Paper presented to Symposium on Man and the Biosphere project, Port Moresby, April, 1975).

White, K.J. (1975). - The effect of natural phenomena on the forest environment. PNG Dept. Forests, Port Moresby. 27 pp.

Womersley, J.S., and J.B. McAdam (1957). - The forests and forest conditions in the territories of Papua and New Guinea. T.P.N.G. Forest Service (Brit. Comm. For. Conf. in Aust., 1957). Govt. printer, Port Moresby.

Womersley, J.S. (1958). - The Araucaria forests of New Guinea: a unique vegetation type in Malaysia. Proc. Symp. Humid Tropics vegetation, Tjiawi (Indonesia), Dec. 1958. UNESCO: 252 - 254.

Some research projects currently being undertaken in Papua New Guinea:

Julian Ash, Research School of Pacific Studies, Australian National University: The ecology of montane forests. Work to date has centred largely on *Nothofagus* forests on the S.E. slopes of Mt. Giluwe (see Ash (1975)).

Niel Enwright, Research School of Pacific Studies, Australian Nationa University: Ecology of *Araucaria* spp. Ph.D. research project. Working currently in the Wau Bulolo area.

Richard Corlett, Research School of Pacific Studies, Australian National University: Ecology of timberline species. Ph.D. research project. Work on Mt. Wilhelm, Mt. Giluwe, and other areas.

Jeremy Smith, University of New England, Armidale: Recently returned from further studies of the non forest vegetation on Mt. Wilhelm (see Smith, J.M.B. (1974)).

The Division of Botany, Lae: Ecology and distribution of *Nothofagus* forests (Nigel Clunie). Aquatic vegetation (Barry Conn: assessment for the Purari hydroelectric scheme).

Department of Forests, October, 1975.

INTERACTION OF PRODUCTION FORESTRY WITH CONSERVATION OBJECTIVES AND MULTIPLE USE OF FOREST LAND IN PAPUA NEW GUINEA

by

John Davidson
Forest Research Station, Department of Forests, Bulolo
(now P.N.G. University of Technology, Lae)

Abstract

The implications of increasing population and effects of natural phenomena on the forests of Papua New Guinea are considered. The physical, physiochemical and sociological influences of forests on the ecosystem are described briefly. An introduction to production forestry in the country is followed by an indiction of future trends in utilization of the forest resource. As an example of several large-scale operations contemplated, the Gogol wood-chip project is described in detail to illustrate use of land classification and co-ordinated land use planning prior to development. Some illustrations are given concerning management plans to meet production, conservation, ecological and multiple use objectives which contribute to a lessening of the social and environmental impact in the Gogol Timber Area. Finally, some recommendations are made concerning future management of large-scale production forestry with conservation and multiple-use of forest land.

My remarks herein are to be viewed as personal opinions which do not necessarily reflect Government policy on development and conservation.

INTRODUCTION

Man must interfere with his environment in order to live, by baring the soil and sacrificing other forms of life. He is an integral component of the environment and must maintain a special pattern of vegetation on various parts of this Earth-whether it be forest plantations, rubber or coffee plantations, pastures or gardens. He also grows or hunts animals for food. As human populations increase and technology becomes more advanced, man's interference with Nature attains greater scope and speed. The inevitability of this interference with nature is the first ecological principle which the world must face.

When man disturbs the balance of nature, other changes follow. These changes may be large or small, short-term or longlasting, predictable or unforeseen, controllable or uncontrollable. A small change in the delicately balanced factors of the ecosystem at one place may trigger off much greater changes in the environment elsewhere. This effect is often termed an "ecological explosion". In this case the explosion is noiseless and often delayed but can be just as devastating.

When this ecological chain reaction begins, man has, or has had, two alternative choices:

(i) To go away and live somewhere else, or
(ii) Stay in the one spot and try and control it.

The first alternative is an escapism practised by the early inhabitants of Papua New Guinea (and elsewhere), in the form of shifting agriculture. This technique of subsistence farming has been responsible for creation of much of the grassland now found in Papua New Guinea and continues to create many tens of thousands of hectares each year. The second alternative will be eventually inescapable as the area of land is limited.

Quite a long time ago (7-20 thousand years; Golson, 1968), in Papua New Guinea there was a delicate balance between relatively small groups of people, the native flora and fauna and the environment. People are larger animals lived a nomadic life to obtain food over wide areas, impingeing only lightly on the total natural resource.

More recently, increased population, with its attendant clearing and burning led to a relationship which ensured maximum productivity for a short time on a given site. There has followed a series of ecological explosions with different environments reacting in different ways depending on their inherent weakness when subjected to stress.

The effects of natural phenomena also on the Forests of Papua New Guinea have been estimated by White (1975). He concludes that the effects of earthquakes, volcanoes, wind, erosion and river systems, fire and so on have kept much of the forest in a late secondary stage and that a stable, climax forest is possibly a rare occurrence in Papua New Guinea.

These points have important implications to be considered in relation to conservation, management and productivity of the forests of Papua New Guinea. In this paper I propose to deal with the influence of forests on the ecosystem and a short history of production forestry, then to describe one of the present major operations in forestry and the land use planning involved, and finally, to discuss multiple use and conservation of forests generally.

FOREST INFLUENCES

Forests have three main influences:-

(i) Physical
(ii) Physiochemical
(iii) Sociological (Cultural and Psychological)

Firstly, forests affect local climate by acting as windscreens, reducing temperature extremes, increasing infiltration and deep seepage of water, regulating stream flow and, in general, maintaining balance within the hydrological cycle. These principles are well established elsewhere if not in Papua New Guinea.

Secondly, the physiochemical influences of forests also are of great importance. In forest vegetation, especially in rainforest, where trees grow close together or on top of one another in layers, there is a very efficient circulation of nutrients between litter fall and litter decomposition through shallow feeding roots back into the trunk, branches and leaves. Very little of this storehouse of nutrients is lost from the cycle under undisturbed natural conditions. Thus, <u>much of the fertility of a tropical forest system is above ground</u>.

Microbiological activities in the humus layers also are of great importance in developing a structure which ensures aeration and permability of the soil. Various species of the micro-flora also fix nitrogen and assist in the solubilization and up-take of nutrients by trees. These natural processes are probably the mechanism of restoration of soil fertility during the fallow period allowed in the system of shifting cultivation practiced within the forests of Papua New Guinea.

If one understands some of this ecological background of the forest, one can imagine what happens when the forest is cut down and burnt. All the complex associations within the forest are broken. The micro-climate is changed, soil fertility declines, changes and nitrogen (particularly) is lost from the system.

Thirdly, the forest has been culturally and psychologically significant in the lives of the rural population in Papua New Guinea. Not only has it been a rich source of a tremendous variety of items in the categories of food, clothing, building materials, weapons, decorations and medicines but it has also been a source of beauty, wonder and fear in a primitive society influenced by sorcery, poison, magic potions, and spirits.

It is most difficult to evaluate the influence of forests in terms of money. A more realistic method to assess and demonstrate the benefits of forests on local water relations and soil fertility is to observe the catastrophe that occurs when they are removed. Flash floods, soil erosion, sedimentation of streams and rivers, creation of marshland and swamps, losses of crops, buildings, animal and human lives are just some of the problems which might be experienced in villages and rural areas. To avoid all these problems is worth a good deal of money, although the money will not be in any particular person's pocket in the way of a normal profit is gained from a natural resource.

PRODUCTION FORESTRY IN PAPUA NEW GUINEA

Forest is defined as a community of trees which form a canopy of crowns that touch or overlap. Thirteen types of forest have been mapped and investigated: five on plains and alluvial fans, seven on hills and mountains and one restricted to south-west Papua New Guinea (Department of Forests 1973; Paijmans, 1975) (Table 1).

On the plains and alluvial fans, differences in the structure and floristic composition are related mainly to drainage conditions and soil type. These forests are found mostly on the coastal lowlands and further inland along the major rivers. This type of forest is often referred to as a lowland rainforest.

Variability in the forests of the slopes and mountains are mainly related to variations in altitude with associated changes in temperature and precipitation. These types grade into one another and they also vary in altitude from place to place due to local effects of topography, soil and climate.

Lower montane and montane forest is characterized by dense, small-crowned trees.

The land mass of Papua New Guinea contains approximately 46 million hectares of which 40 million (87%) are covered with forest. The remaining area is classified as swamps, waters, grasslands and commercially unproductive savannah forests. The potential of the forest to assist in the economic development of this country is considerable as Papua New Guinea has the second largest forest area in the Pacific (Fig. 1).

Approximately half of the forest area in Papua New Guinea is located on broken and mountainous terrain, far away from roads (Fig.2). At present and foreseeable levels of technology and utilization, the productivity of these areas is severely limited by virtue of poor accessibility.

Until 1940, the impact of forests for timber was of no significance compared with the area converted to agricultural plantations. The expatriate population was small and its needs were supplied from imports or small local sawmills.

Exports were non-existent until near the beginning of the second world war when a small log export trade of some 7-8 thousand cubic metres per annum was built up. (White, 1975a).

The war years stimulated a local demand for timber and this demand increased rapidly during the post-war reconstruction and development programme. The log volumes harvested rose from 46,000 cubic metres in 1952 to over 800,000 cubic metres from 1972 onwards. The rapid increase in harvest occurred after 1963 (Fig. 3) and over half was exported. Almost all of the harvest came from the lowland rainforest because of its accessibility.

The pre-1963 demands on the forest had a gentle impact due to the selective type of logging, in which a few valuable species of large size were harvested. Some of the harvested areas were converted to forest plantations of quite small scale on world standards (Fig. 4). Other areas were used for agriculture and small areas were abandoned after logging. The continuation of such a regime with perhaps improved forest management practices in logged-over areas would not have been a cause of concern to environmentalists or conservationists. However, this situation was not to continue.

The Government of Australia requested an economic survey mission by the International Bank for Reconstruction and Development (I.B.R.D.) to undertake a review of the economic potentialities of Papua New Guinea. It was to assist in planning a development programme designed to expand and stimulate the economy and thereby raise the standard of living and enable the nation to be self-reliant economically after Independence (I.B.R.D., 1965).

The mission, in a concise recommendation on forestry, stated "Forest products are considered capable of being stepped up from approximately 200,000 cubic metres of log cut in 1962/63 to a level of at least 750,000 cubic metres annually in five years. This rate of output could be easily maintained there-after, and in fact, increased without depletion of forest resources. Large-scale enterprise will be needed for the logging and processing of timber".

The Australian Government accepted the recommendations of the I.B.R.D. report and, through the Department of Forests, has promoted the development of large-scale export-orientated industries. These, of necessity, must be based on large resources in limited areas. Of the inventoried resources (Fig. 5) those given in Table 2 have, so far, been selected for promotion and active development. These are new industries, additional to some 90 sawmills supplying the domestic market and obtaining their logs from some 800,000 ha. of forest.

THE FUTURE OF PRODUCTION FORESTRY IN PAPUA NEW GUINEA

Pressure on the forests of Papua New Guinea will continue to increase as a result of a desire for social and economic development. Forest utilization will continue to exert a major and increasing impact because:-

1. *it is considered that the promotion of forest industries is in accord with the Government eight-point national programme (especially points 3,4,5 and 6) (see Appendix 1)*

2. *the market acceptance of Papua New Guinea timbers continues to grow. This trend is well illustrated in the log export figures for the period 1951-1974. (Table 3).*

3. *some of the present projects have adjacent large areas of forests potentially available to them and, apart from those resources so far mentioned, there are additional large areas available (Fig. 5).*

4. *recent technological developments have enabled the use of mixed lowland rainforest as a source of wood chips for pulp.*

5. *a world-wide shortage of wood raw material, especially for pulping, has occurred. A recent F.A.O. report states "The world cumulative 1974-1978 paper grade pulp deficit is estimated to be 17,000,000 tons of which 4,500,000 tons are groundwood, 1,000,000 tons semi-chemical and 11,500,000 tons chemical pulps." "The world cumulative 1974-1978 paper deficity is estimated to be 8,000,000 tons." "The fibre resource physically most readily available is the tropical forests."*

How will these pressures affect forestry operations in Papua New Guinea? The large scale of operations will necessitate full forest harvesting over large tracts of forest situated close to ports and infrastructure. This development will require the preparation and implementation of detailed land use plans and the application of intensive forest management plans.

Chip projects are planned in the major development areas mentioned in Table 2. Pulp mills are being considered.

One chip project is operational in Papua New Guinea. This is at Gogol, near Madang. I will use this project as an example to show that co-ordinated land use planning will be essential to ensure that, in future development programmes, there will be a realistic implementation of utilization and conservation practices.

THE GOGOL WOOD CHIP PROJECT

i) Location and Physical Features:

Gogol Valley is located to the south and west of Madang at a distance of 20-50 kilometers. The main road to the logging operation crosses the Gogol River at Mawan, some 40 kilometers from Madang.

The Valley is extremely low-lying throughout. There is a low saddle 150 metres above sea level (m.a.s.l.) dividing the headwaters of the Gogol system from the Sogeram and Ramu River systems. The Gogol River also drains parts of the high mountainous country of the Adelbert Range and the north western tip of the Finisterre Range. Some 50 kilometers from the coast, the floor of the Gogol Valley is only about 70 m.a.s.l.

In relatively recent geological time a major fault developed up the Gogol Valley between the coastal limestone scarp (which extends from Amele to the Adelbert Range in the east) and the limestone foothills of the Finisterre Range in the west. The fault extends into the Ramu-Markham geological fault.

Earth tremors have caused considerable damage to the landscape throughout the area in recent times. The severest seismic disturbances have occurred in the 1910's, early 1930's and early 1970's. Scars of old landslips are a noticeable feature of the steeper slopes of the area. A high percentage of these slopes appear to have been denuded at some time in the past. The type and magnitude of damage is illustrated by a recent severe earthquake centred about 60 kilometres north-west of Mawan which completely bared 80 percent of the land surface over an area of 500 square kilometers outside of the area of the timber project.

During major droughts fires have caused extensive damage from time to time. Probably most of the area has been subjected to fire at least once. In living memory fires have occurred in the early 1900's, around 1915 and the last major one in 1941.

ii) Soils:

The soils of the Gogol-Nuru area have been considered for a long time unsuitable for intensive cropping. Most of the soils are poor in nutrients and have been intensively leached except for the banks of the rivers which are subjected to regular flooding and siltation.

A broad-scale land system map has been prepared by C.S.I.R.O. Division of Land Research (unpublished Divisional report 57/2) and the terrace soils in the Gogol Valley have been described; most are deep alluvial soils, and textures range from sands to heavy clays. The soils in Baku area range from deep, welldrained ones to some with variable textured profiles, consisting of layers of coarse and fine material to considerable depth with strong gleying throughout. Flooding is common in some parts and, even on sites not usually flooded, water tables are commonly at or near the surface during the wet season. The water seems to be held by impermeable sedimentary strata.

The subsoil on the slopes is often a mottled plastic clay which tends to slip if the layers are disturbed for road construction. Movement in these clay soils is considerable during the processes of wetting and drying.

iii) Climate:

The climate is monsoonal. Heaviest rainfall occurs from December to April inclusive. The dry season during the period June to November coincides with the south-east trade winds which are partially deflected by the Finisterre Range.

Rainfall is around 2,300 mm, falling during 100-120 wet days (greater than 5 mm/day) per annum.

iv) Vegetation:

Existing vegetation patterns are undoubtedly the result of past land use by man with additional influences of natural phenomena such as the fires, floods and earthquakes already mentioned. Much of the area is covered with regrowth forest and a mosaic of grassland through the full age spectrum from recent gardens to older regrowth approaching a rainforest type.

Most areas near old villages have become permanent grasslands which are regularly burnt over each dry season destroying any regeneration of tree species.

Major forest fires have altered the species composition of the forest area, favouring the more fire resistant genera such as Intsia at the expense of less resistant genera such as Pometia.

The existing vegetation is in a state of flux. It is extremely doubtful if areas of tropical lowland rainforest in a climax condition could be found except, perhaps, in protected spots along the larger creeks.

v) Development of a Forest Industry:

At the beginning of this century, the Administration examined the Gogol area for economic development. A large tract of flat land was surveyed and acquired. Owing to shortage of labour, the proposed economic development failed to materialise.

In 1930's the land was examined again for possible plantation development but proposals were shelved.

When looked at further in the 1950's, the area was considered unsuitable for intensive agricultural cropping.

The Government, in 1951, sought a suitable forest area containing an adequate resource to supply Madang with sawn timber, as considerable quantities were required for reconstruction of the town and plantation buildings. A survey was undertaken of the area acquired early in the century but the forest resources were considered inadequate for sawmilling and uneconomic to log because of poor access, limited volumes per hectare and long distance from the town.

Interest was revived again in 1961 when a forest area was sought to aid economic development in the Madang area. An area further up stream of the original acquisition was surveyed and forests containing marginal volumes per hectare of suitable species were located. The merchantable trees were usually located in gallery forests.

A detailed survey was made subsequently proving a timber resource covering about 52,265 hectares. A timber-rights lease of 40 years duration was acquired over this area from the owners in 1963. By this time there was a poor road to the east bank of the Gogol River.

The right to harvest logs from the area was offered to public tender and widely advertised in 1964. Only one applicant was received. Some time later this application was withdrawn. A further offer of tender in 1966 produced the same result; one application, later withdrawn. The Government then called for development proposals with a closing date of 31 August, 1968.

Meanwhile, the Catholic Mission at Utu began operating a small bush sawmill near Utu in the Gogol Timber Area in 1965. The Mission selectively logged species required for their own construction programme and were allowed to make limited sales of sawn timber.

Only one development proposal was eventually received from a consortium of Japanese Companies. This consortium could not satisfy foreign investment guidelines (that an Australian or Papua New Guinea partner hold at least 51 percent of the equity capital), so the negotiations broke down in December, 1968. However, a few weeks later, the investment rules were relaxed and the Government recommenced negotiations with the same consortium which subsequently became Jant Pty. Ltd.

During the short break in negotiations with the Japanese consortium, a local company, Wewak Timbers Pty. Ltd., applied for the right to harvest the Gogol area, so they entered the negotiations. At the same time, the consortium found that sawn timber of Intsia spp. was extremely difficult to sell in Japan and brought very low prices. They re-examined the Gogol forest resource to find that, excluding the Intsia spp., the forest was suitable for a wood chip operation even though prices for wood chips were depressed at that time. Their application was modified to undertake veneer and wood-chip production, thus enabling Wewak Timbers Pty. Ltd., to undertake the sawmilling components. The latter established a small bush mill in the Gogol Timber area in May, 1969 and selectively harvested immediately adjacent areas.

Negotiations continued until August, 1971 when an agreement was signed to enable the Japanese consortium to harvest the area for wood chips and veneer logs and to supply saw-logs to Wewak Timbers Pty. Ltd.

A permit was issued to Jant Pty. Ltd. in July, 1972 and land lease over the industrial site in Madang was issued in May, 1973. At this time Wewak Timbers Pty. Ltd. moved its mill to Madang.

The chip mill was constructed between May, 1973 and March, 1974. The final long-term timber permit was issued to Jant Pty. Ltd. in March, 1974, some ten years after the timber rights were purchased. The first ship-load of chips was despatched to Japan in June, 1974.

vi) Land Classification:

Land classification has been carried out in the Gogol Timber Area based on the assumption that areas of uniform natural vegetation type and uniform terrain type are necessarily of uniform suitability for a particular type of land use. The technique is a variant of the "landscape" approach used widely in resource surveys and integrated land use planning overseas. This approach has several advantages in application to the Papua New Guinea situation:

a) it introduces a classification which is amenable to mapping;

b) it requires no assumption regarding the type of end use contemplated;

c) it permits maximum use of aerial photography in its compilation;

d) it is simple in concept; and

e) it is generally applicable to similar project areas elsewhere in Papua New Guinea.

Land variation in the Gogol Valley and elsewhere can be understood through its effect on vegetation. In turn, vegetation types can be defined in terms of the effect of four variable, relatively independent, factors on the "climax" vegetation of the area. The inter-relationships among thirteen recognised vegetation types, which are by definition broadly equated to land types, and the four variables: flood susceptibility, slope, drainage and impact of man, are given in Fig. 6 for the Gogol Timber Area.

Having defined the land types, it was possible to ascertain the development potential of the individual land types for forestry purposes (Table 4).

vii) <u>Land Use Planning</u>:

The Government set up a Timber Working Committee to monitor the Gogol operation, to offer advice to the people, the company and the Government, and to prepare a land use plan. This committee consists of representatives from involved Government Departments, Villagers and Landowners.

Forestry development was, therefore, integrated at an early stage with agricultural and other forms of development in the Gogol Timber Area.

A major part of the area was required for reforestation to provide a resource for industry in perpetuity. This aspect dominated land-use planning and involved the following:-

a) Classification of the timber area into "land types" which are internally homogenous in terms of suitability for commercial reforestation (derived from the 13 land/vegetation types previously determined):

b) evaluation of individual land types in terms of suitability for commercial reforestation, using a five-point scale;

c) estimation of the total area of forest plantations required to supply the Madang Timber Industry indefinitely;

d) demarcation of that area, moving from the best available site types to progressively poorer types, with allowances for:-

1. requirements of land for subsistence farming in localised areas of high population pressure
2. requirements for small holder agricultural and cattle pursuits
3. requirements for large-scale agricultural and pasture development
4. requirements for reserves with no disturbance whatsoever allowed and
5. localised unwillingness of owners to lease their land.

After taking into consideration as many factors as possible, the Timber Working Committee prepared a number of scenarios in an effort to gain the most suitable use of land.

The scenario finally chosen directed the land area into productive categories for forestry and agriculture, various types of reserves (environmental reserves, benchmark reserves, village reserves) and inaccessible wilderness areas. The productive areas were further subdivided into areas suitable for intensive agricultural cropping, grazing, intensive forest plantations and less intensively regenerated forest.

viii) Social and Environmental Impact:

This type of development meets the desires of the Government as expressed in the eight-point plan. It is a decentralised operation and employs people from the locality. It will greatly improve the wealth of the area as the local people benefit from the profits in that they will own a major share of the reforestation company and eventually own a part of the total operation.

However, the large scale of the operation will have a considerable sociological impact on most of the people living in the Gogol Valley.

Already, most of the main villages have been linked with the main road and the remainder will have similar access within the next five years. Apart from income directly from shares in Jant Pty. Ltd. and associated Companies, there will be ample opportunities for business operations on a community basis such as contracting, farming, transport operation and trading, and as employees of these entrepreneural activities.

The timber industry in Madang at present employs some 550 people and employment is expected to reach a peak of about 1800 in 1977.

The land belongs to the people. It will be intensively developed over the next 10 years with Jant Pty. Ltd. spending some K10 millon on roads and bridges alone. The owners will lease part of their land to the reforestation company and will involve themselves in the development of agricultural and pastoral pursuits on the remainder of the area cleared by harvesting. The Gogol Valley has limited potential apart from forest development, therefore, the greatest financial return for labour and material input will be for forest plantations and perhaps grazing as a second industry. There should be a minimum of pressures from people desiring other avenues of production, in the long term, on much of the area to be harvested over the next 20 years.

People of Madang Province will come to depend more and more on the timber operation to satisfy their rising expectations for employment and improved life style. They will come to see the harvesting of the existing forest as the start of a cycle which will enable their land to be productive in perpetuity. In due course, the operational control will pass to them and they will become conscious of the value of this operation to the community and to the nation. As the people become more aware of this, they will ensure that adequate land is made available for reforestation so that the chip-mill will be upgraded, firstly to a pulp mill and, eventually, when adequate plantation supplies of wood are available, perhaps to a fully integrated paper mill.

It was realised at an early stage that little is known of future effects of the almost total harvesting of a tropical lowland rainforest located on sedimentary clays within the precipitation pattern of the Gogol area. It was appreciated that there would be higher rates of run-off and erosion immediately after logging. The early logging ran into many problems such as land-slides, road slips, washouts, snig-track gulleying and marshy ground becoming waterlogged.

Most of these problems have now been overcome by the vigorous application of a management plan which takes into account production objectives, ecology and conservation of the enironment while allowing multiple use of the land.

ix) Management Plans to Meet Production, Conservation, Ecological and Multiple-use Objectives in the Gogol Timber Area.

Both the Government and Jant Pty. Ltd. are keen to ensure that the landscape environment is impacted in such a way that it will recover or maintain its equilibrium so that it will provide for the organisms dependent upon it in perpetuity. The main difference between the alteration of the vegetation in this operation and the alteration by man and natural agencies in the past is that man's needs on a continuing basis have been taken into account.

The Management of the Gogol Timber Area is based on a five-year working plan which is updated annually. The plan, apart from simply ensuring an adequate supply of raw material to the industry, considers such matters as management of logging to ensure:

a) land is available for intensive reforestation at the best time for planting

b) protection of the environment

c) conservation of as much of the natural landscape as possible and

d) multiple use of land wherever possible.

In each annual up-dating of the five-year plan, greater attention and detail is given to the operations for the forth-coming year. The following items are among those specified:

a) size of logging area and estimated volume of logs to be extracted

b) location of roads and major snig tracks

c) location of sites for log-loading ramps

d) location of green breaks for fire control, maintenance of water quality and for aesthetic purposes

e) demarcation of village reserves.

In determining the extent and the part of the resource to be harvested, areas are designated which, due to steepness of slope and unsuitable soils, are prone to erosion. Road and snig-tracks are designed to ensure:

a) erosion is kept to a minimum by installation of cross drainage and water dispersal mechanisms allowing free movement of water especially on steep slopes.

b) ponding effects are avoided on flat country

c) that, where large amounts of loose fill result, these are compacted and side-slopes rehabilitated with suitable vegetation

d) roadside verges are revegetated.

Trials are underway using legumes and grasses for revegetation and stabilization of exposed soil. Indications are that these areas can be covered rapidly.

Logging is not generally carried out within 20 metres of the bank of a permanently flowing watercourse. These strips of vegetation are important in the preservation of drainage patterns and water quality and for fire protection. Logs are not hauled through the green breaks or flowing streams unless absolutely essential. Should it prove necessary, the shortest possible route is taken and care exercised to protect the green break and watercourse. Any debris of logging operations which impedes a creek is removed.

Monitering of turbidity in streams during and after heavy falls of rain is in progress. A number of stream gauging weirs have been established at the outlets of small catchments to moniter discharge before, during and after logging and reforestation activities.

As logging ramps are a major source of pollution, they are kept to the smallest possible size and debris is disposed of to reduce insect problems and present a more attractive appearance. To alleviate the severe compaction of logging ramps and prevent ponding of water, deep ripping is carried out followed by sowing a cover crop after logging is finished.

Village reserves are established annually, while food trees and selected seed trees are marked for retention on a continuing basis.

Logging commenced for chipwood material December, 1973. It is now possible to see the effects of total harvesting on the landscape over a period of almost two years. Most areas have attained a complete canopy 3-4 months after harvesting. The vegetation is 1-2 metres high after 6 months. To date, there is no evidence of any previously forested area becoming grassland. With a heavy ground mat of rotting debris which will largely disappear over a period of 2-3 years, there is little danger of erosion or of a major fire.

Some swampy areas harvested during the wet season are slow to regenerate owing to the presence of free surface water. These areas are under study at present.

Stringent pollution controls were enforced for the industrial site at Madang, to ensure that there would be a minimum of detrimental effect on the environment. Controls were placed on noise levels, airborne dust, ash and fumes and on purity of water discharged from the site. Suitable machinery was installed to meet these requirements.

These management plans are not static, but are continually revised as new information becomes available to either the Company or the Government. By continuing co-operation between all parties involved in the Gogol Timber Project, it should be possible, in the future, to present an operation which will be a model of its kind.

CONSERVATION AND MULTIPLE USE OF FORESTS

Major factors leading to alteration of the forest ecosystem are recognised and some have been referred to here.

It is likely that most disturbance outside light, shifting cultivation will move the floristic composition away from the climax desired by most conservationists.

For restoration to be contemplated under these conditions it is desirable to determine, beforehand, what the maximum alteration to the ecosystem is likely to be an then consider whether restoration will be possible or worthwhile.

Because of the pressures coming to bear on the forests and the difficulties associated with long term management of a natural ecosystem, the following points should be considered:

a) it is vital to classify now and retain representative areas of forest throughout Papua New Guinea.

b) it will be necessary, for each of these reserves, to determine the minimum area necessary to allow the ecosystem in that place to retain its identity.

c) apart from reserves, the use and development of the forest resource should be planned to give a balance between short and long-term objectives, that is, to ensure the forests contribute substantially to national development now and continue to do so in the futher.

d) it is highly desirable that the Government exercise strong control over the rate, and ways, the forest is utilized.

e) extensive clearing of forest, either for intensive forest production, or agricultural pursuits, should be approved only on the basis of sound land-use planning.

f) standards of management planning and environmental control should be defined and their implementation enforced by legislation.

g) the Government should ensure public funds are made available to meet increasing investment in suitable facilities and equipment and an increasing expenditure on administration and research associated with multiple use of forest, so that the timber industry does not subsidize the rest of the community in the the provision of these goods and services.

h) organizations, clans or individuals controlling sizeable areas of forest land, more or less in its natural state, should ensure operations over which they have control never lead to regional elimination of any species of wildlife. Sites of particular scientific significance should be preserved.

i) there should be continual evaluation of logging equipment and logging systems with respect to their effect on the environment. An early preparation and publication is desirable of an illustrated guide to environmentally sound logging practice under Papua New Guinea conditions.

j) the Forest Authority or the forest industry should employ the services of at least one professionally trained officer with the ability to recognise wildlife problems and advise on wildlife matters.

k) funds should be made available to support continual research into aspects of forest ecology.

l) a survey should be undertaken to identify forms of forest use other than wood production and agriculture, particularly the capacity of forest areas for various forms of recreation.

m) the forest authority and forest managers need to seek the advice of trained landscape architects and the services of other appropriately trained specialists in order to devise effective and imaginative programmes of forest development.

n) there should be a means of obtaining advice and views of the public in the manner in which their forests should be managed and developed.

o) the training programmes for foresters in Papua New Guinea should contain appropriate courses dealing with such topics as ecology, the environment, pollution, sociology, landscape architecture and the interaction of these with production forestry.

This list is not exhaustive. For most of those points requiring legislation or enforcement, a mechanism for achieving this perhaps already exists in the National Investment Priorities Schedule (Appendix II), particularly point 7 of the minimum terms and conditions, and existing forest policy (Department of Forests, 1974). From the point of view of education, the Degree in Forestry at the University of Papua New Guinea and Papua New Guinea University of Technology emphasises many of the points mentioned, however there is a need to have conservation education in primary and high schools, and, elsewhere, to ensure training in environmental subjects reaches as large an audience as possible.

CONCLUSION

This country must avoid the syndrome which pervaded the early history of many of the developed countries. This was, if a man cut down one or two trees he ran the risk of being called a vandal; if he cut down a few hundred trees he may have become a successful timber man; if he destroyed millions of trees in opening up the land for development, he was acclaimed a pioneer and he may or may not have become a successful farmer or businessman as well.

In the interests of national development some pioneering is necessary, as in the following verse-

"I split the rock;
I felled the tree;
The nation was -
Because of me ....

Mary Gilmore.

However, on the broad scale, one must keep the ecological balance between development and preservation, which means conservation. To do this, firstly, we need a balanced definition and use of habitat reserves for inspiration and biological reference and, secondly, to ensure that development of our forests is carried out with the environment protected and conserved.

ACKNOWLEDGEMENTS

The background information on the Gogol Woodchip project was compiled mainly from Cavanaugh (1975), while some of the aspects on management, ecology and environment in forestry activities is drawn from White (1974, 1974a, 1975, 1975a), who has written extensively on these and related aspects, particularly in regard to the lowland rainforest. Stastical information was obtained from annual reports of the Department of Forests and miscellaneous reports of the Food and Agriculture Organization. The land classification for the Gogol area is largely the work of Mr. Whyte of the Department of Forests. The final report of Expert Panel No. 1: Ecological effects of increasing human activities on tropical and subtropical forest ecosystems in the rogramme on Man and the Biosphere of UNESCO is recommended for background reading as are the proceedings of the Symposium of Man and the Biosphere Project dealing with ecological effects of increasing human activities on tropical forest ecosystems, held in Port Moresby in April, 1975.

REFERENCES

CAVANAUGH, L.G. (1975). The Gogol Woodchip Project. Symposium on Man and the Biosphere Project, Port Moresby.

DEPARTMENT OF FORESTS (1973). New Horizons. Jacaranda Press, Brisbane.

DEPARTMENT OF FORESTS (1974). Forest Policy.

GOLSON, J. (1968) Archaeological prospects for Melanesia. In I. Yawata and Y.H. Sinoto (eds.), Prehistoric Culture in Oceania; a Symposium, Honolulu, Bishop Museum: 3 - 19.

INTERNATIONAL BANK FOR RECONSTRUCTION AND DEVELOPMENT (1965). The economic development of the Territory of Papua and New Guinea. John Hopkins Press.

PAIJMANS, K. (1975). Explanatory notes on the vegetation map of Papua New Guinea. C.S.I.R.O. Land Research Series No. 35.

WHITE, K.J. (1974). Political and social demand and impact on the rainforest. UNESCO Sub-Regional Seminar on Ecological and Environmental Sciences, Bogor, Indonesia.

WHITE, K.J. (1975). Effect of natural phenomena in Papua New Guinea Forests. Presidental address. P.N.G. Scientific Society, Port Moresby.

WHITE, K.J. (1975a). Planned development in lowland rainforest in Papua New Guinea, with special reference to infra-structure, mining and forestry. Symposium on Man and Biosphere Project, Port Moresby.

WHITE, K.J. (1975a). Planned development in lowland rainforest in Papua New Guinea, with special reference to infrastructure, mining and forestry. Symposium on Man and Biosphere Pfoject, Port Moresby.

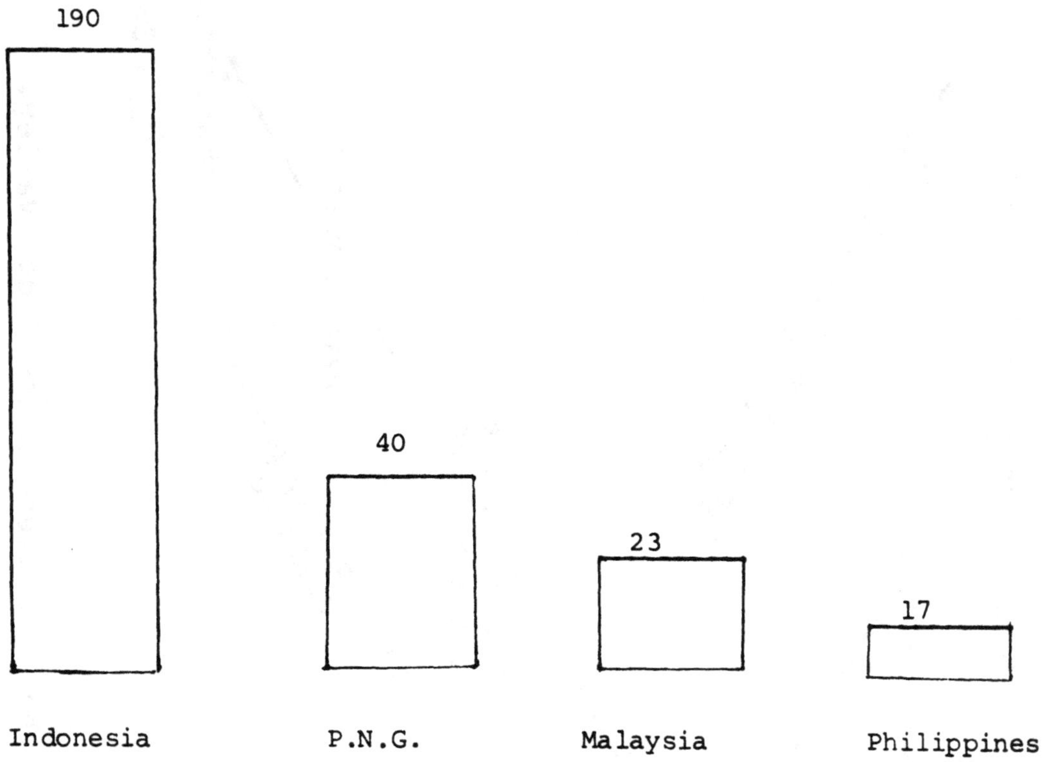

FIG. 1: Gross forst area comparison for countries in the west Pacific (figures are millisons of hectares). (Source: National Forest Policy of Papaua New Guinea, 1974).

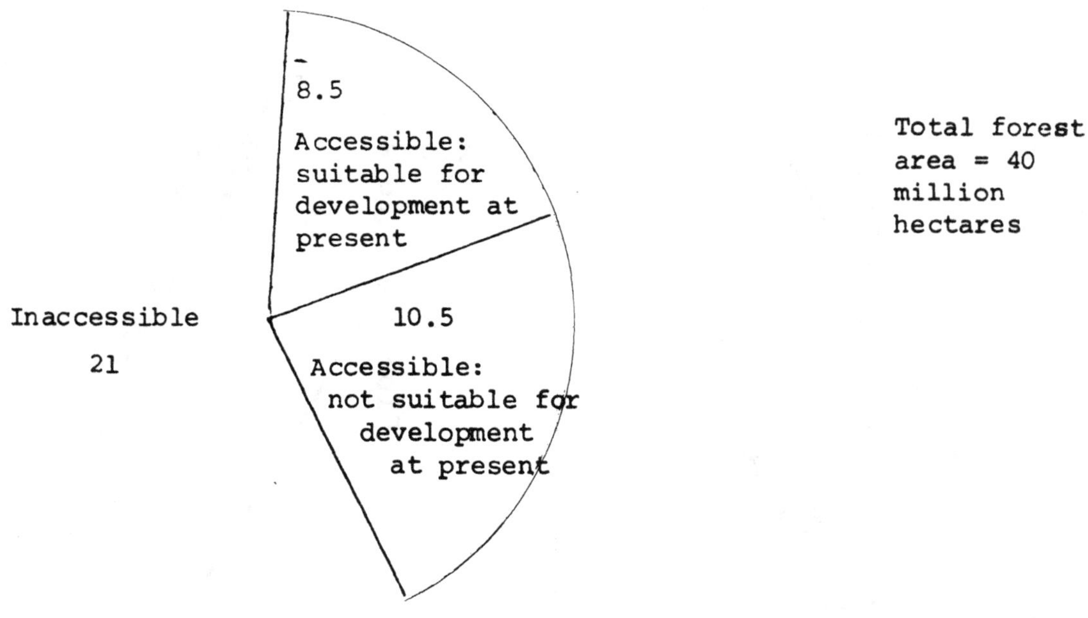

FIG. 2 Accessibility classification of forest areas in Papua New Guinea (Source: National Forest Policy of Papua New Guinea, 1974).

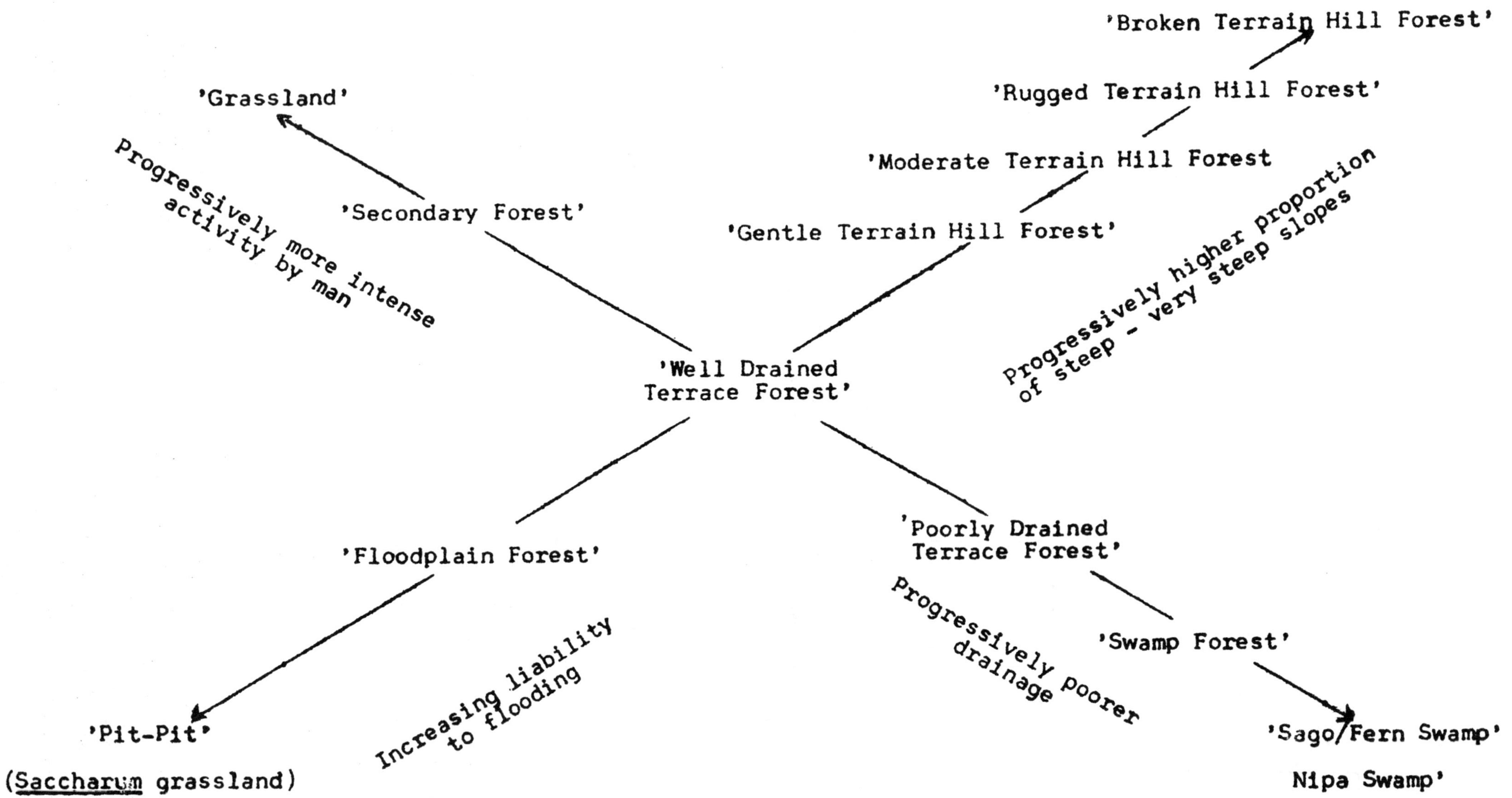

FIG.6: Thirteen vegetation types related to four factors of the ecosystem in the Gogol Valley.

Source: Forest Activities Gogol,Valley Madang, an information paper prepared for the visit of members of the Australian Standing Committee on Forestry, April, 1974.

TABEL I: TYPES OF FOREST IN PAPUA NEW GUINEA

Source: Paijmans, 1975.

Forest on plains and fans	Large-to medium-crowned forest Open forest Small-crowned forest Littoral forest Swamp forest
Forest on hills and mountains Lowland hill forest zone (Sea level to 1400 m)	Medium-crowned forest Small-crowned forest Large crowned forest
Lower montane forest zone (1400-3400m)	Lower montane forest, general type Coniferous lower montane forest Very small crowned lower montane forest
Montane forest zone (3400 m to the forest limit at 3900 m)	Montane forest
Forest restricted to south-west Papua New Guinea	Dry evergreen forest

TABLE 2: FOREST AREAS SELECTED FOR PROMOTION AND ACTIVE DEVELOPMENT

Location	Date Advertised	Taken Up	Total Concession Area. Hectares
Gogol	September 1964	1971	67,000
Sagarai-Gadaisu	January 1971	1975	120,000
Kumusi	February 1971	1975	85,000
Open Bay	August 1970	1973	183,000
Kapiura	June 1973	1975	83,000
Kapuluk	January 1974	1975	181,000
Vanimo	1968		278,000

TABLE 3: LOG. EXPORT FIGURES FOR THE PERIOD 1951 TO 1974

Period	Ann. Log Ex ports m^3
1951-62	4,000
1963-66	44,000
1967-70	135,000
1971-74	410,000

TABLE 4: FORESTRY POTENTIAL OF SOME MAJOR GOGOL LAND TYPES

Note: Areas shown here are to be taken as a broad guide only.
Source: Fores t Activities Gogol Valley Madang, an information paper prepared for the visit of member s of the Australian Standing Committee on Forestry, April, 1974.

Land Types	Volume per acre (cubic metres/ha) Saw	Chip & Saw	Major Species making up 50% total volume.	% Impt. (Vol)	Logability	Estimated Suitability for reforestation	Area (ha)
Well Drained			Intsia	12.01			
Terrace Forest			Terminalia	5.78	Good	Very high	3,500
			Teysmanniodendron	4.78			
Poorly Drained							
Terrace Forest	35	104	Celtis spp.	2.97	Some problems	Low	6,500
			Miscellaneous	2.69			
Floodplain Forest			Myristica	2.53	Good	Very high	3,000
			Ficus	2.51			
Swamp Forest			Pometia Pinnata	2.49	Poor	Very low	500
			Maniltoa	2.46			
			Elaeocarpus	2.22			
			Sterculia	2.22			
			Diospyros	2.22			
			Spondias	2.20			
			Artocarpus	1.96			
			Eugenia	1.96			
Hill Forest			Intsia	9.13			
Gentle Terrain			Celtis spp.	6.34	Good	High	14,000
			Ficus	5.51			
Hill Forest			Pometia Pinnata	5.14			
Moderate Terrain	32	99	Terminalia	3.61	Good	Low	13,000
			Pimeleodendron	3.42			

TABLE 4 (Cont.)

Hill Forest Rugged Terrain			Dysoxylum Celtis Hard Artocarpus Sterculia Pterocymbium Miscellaneous Spondias	3.07 2.88 2.86 2.69 2.64 2.37 2.13	Good-Poor	Low	8,000
Hill Forest Broken Terrain					Poor	Very Low	2,000
Secondary Forest		17			Good -poor depending on terrain	Very high to very low depending on terrain	8,000
Grassland	-	-	-		-	High to very low depending on terrain	2,000
						TOTAL	60,500

APPENDIX I. NATIONAL EIGHT POINT IMPROVEMENT PROGRAMME TRANSLATED TO A NATIONAL FOREST POLICY

SOURCE: FOREST POLICY PAPUA NEW GUINEA, 1974.

1. A rapid increase in the proportion of the economy under the control of Papua New Guineans.

   Increase opportunities for local participation in forest acitivites and encourage the development of locally owned or predominantly locally owned forest enterprises including reafforestation.

2. Equal distrivution of economic benefits including a move towards equalizing incomes.

   Direct a greater share of the benefits from forest development towards the land owners and other nationals involved in the industry.

3. Decentralized economic activity planning and Government spending, with emphasis on agricultural development and village industry.

   Forestry is rural industry - promote forest industries in rural areas, stimulate village participation - involve forest owners and local government in all phases of forest development, including reforestation and management.

4. An emphasis on small scale artisan services and business activity.

   Create opportunities for local businessmen and artisans to participate in transport, logging and all other acitivities related to forest industry.

5. A more self-reliant economy less dependent for its needs on imported goods and services.

   Prompt the use of local forest products to replace imported goods.

   Replant and manage the forests so that they will continue to provide for the needs of future generations of Papua New Guineans.

6. An increase in capacity for meeting Government spending needs FROM LOCALLY RAISED REVENUE.

   Promote the development of profitable forest industries to improve revenue returns through royalties and taxes.

7. A rapid increase in the active participation of women in both economic and social activities.

   Explore ways of increasing opportunities for women to participate in forest activities.

8. Government control and involvement in those sectors of the industry WH ere control is necessary to achieve desired kind of development

Forests are a national asset and must be managed in the national interest.

To ensure the people of Papua New Guinea both now and in the future receive maximum benefits from the management of the forests the National Government will formulate and regulate policies governing the use of forest resources.

APPENDIX II: NATIONAL INVESTMENT AND DEVELOPMENT GUIDELINES FOR FORESTRY

SOURCE: Post Courier, Wednesday, 23 April, 1975.

Guidelines

1. As a general rule, no agreement between the Government of Papua New Guinea and a foreign enterprise will be by an Act of the House of Assembly;

2. In the appraisal of investment projects a company's records of employer/employee relationship in other countries will be taken into account; and

3. The technology utilised should be the most labour-intensive determined to be economically feasible.

Minimum Terms and Conditions

Any foreign enterprise or individual wishing to engage in an integrated timber industry must agree to:

1. Comply with the laws of Papua New Guinea as amended from time to time;

2. Permit Government or Papua New Guinean equity participation or options of equity participation to the extent determined to be practical and appropriate;

3. Train Papua New Guineans for all levels of employment under a specific programme approved by the Government;

4. Within the framework of the industrial relations legislation, agree to involve to the maximum extent practicable, Papua New Guinea employees in decision making relating to terms and conditions of employment;

5. Encourage and assist Papua New Guinea corporate entities or entrepreneurs in establishing business associated with the timeber projects and directly connected with the main timber operations;

6. Utilise sources of supply and services available in Papua New Guinea, preferably Papua New Guinean owned, unless adequate justification for not doing so can be provided;

HABITAT CONSERVATION IN PAPUA NEW GUINEA

by

R. J. Johns
P.N.G. Forestry College, Bulolo

With increasing population pressure and consequent destruction of the rain forest, particularly for agricultural purposes, it has become important for New Guinea to establish a series of ecologically viable reserves. The present rate of destruction of the forests and grasslands of Papua New Guinea could, within the next 20 to 30 years, mean that many important habitats will no longer be available either for education, research, or recreation. The Department of Forests is under continuous pressure to select and establish reserves in any area for which a major timber development scheme is proposed. Despite the much greater area being destroyed for agricultural development, little pressure is exerted on the agricultural developers to establish similar reserves.

Because of the unique nature of the vegetation of Papua New Guinea, it will be necessary to establish three major types of management systems for any parks in P.N.G. These should provide an adequate series of sites to both preserve the major ecosystems, and to fulfill the social requirements of the people.

THE PARKS

A cursory survey of the vegetation of Papua New Guinea often gives the visitor a misguided impression as to the age and origin of the forests. The bewildering diversity of the plant species in the tropical forests suggests to many people an old vegetation analagous to the 'climax' vegetation, as described by many authors from tropical areas. As detailed ecological studies are conducted in the forests of Papua New Guinea, it has become increasingly apparent that large areas are dominated by secondary forest. Forests dominated by species such as *Anisoptera, Pometia, Intsia, Elaeocarpus, Castanopsis, Eucalyptopsis, Auracaria, Nothofagus,* and *Dacrydium,* are, in many cases, the result of a wide range of catastrophic occurrences. Certain of these (gardening, fire)would occur infrequently in any national park, despite the major modifying role that they have played in the vegetation of Papua New Guinea. Natural disasters such as volcanic activity, landslides and cyclones, can still be expected within park areas. Any policy aimed at the complete protection of areas of secondary forest will result in the gradual development of a mixed rain-forest, although admittedly over long periods of time.

GRASSLANDS AND FOREST MARGINS

Preservation of grassland types and forest margin situations in Papua New Guinea are of extreme importance for the conservation of often unique (and rich) habitats for plants, birds, and animals. The diversity of plant types in the forest margins is of the utmost significance, as it is in this zone that many food plants for a wide variety of animal and bird species occur. Any management systems which result in the destruction of grasslands and consequent destruction of the forest margin situation would result in a marked decrease of species diversity in the park area.

The anthropogenic grasslands of Papua New Guinea themselves deserve preservation. If firing is excluded from such grasslands in any reserved area, the establishment of shrubs (which will grow in a period of 6 to 10 years) will inevitably remove habitats of outstanding biological interest.

PARK ESTABLISHMENT

Three major categories of parks are envisaged. Each will require quite distinctive managerial objectives.

1. PARKS FOR RECREATION AND EDUCATION

Primary emphasis should be placed on the establishment of parks in areas close to all major population centres. The educational value of parks is especially important, as in Papua New Guinea traditional attitudes are quite opposed to concepts of conservation. Traditionally the forests were regarded not as an asset, but as a hindrance, particularly with relation to subsistence agriculture.

It is proposed that the parks when established, should have a measure of financial self-sufficiency. Present laws relating to the preservation of the Bird of Paradise will prove quite inadequate to protect the different species. Native trading in Bird of Paradise plumes occurs, and illegal hunting with guns, combined with habitat destruction, particularly for agricultural development, will result in extinction of many species. The extinction of the Huia in New Zealand, previously a protected bird with immense traditional importance, was a direct result of such a policy.

It is proposed that breeding farms for various birds, including the Bird of Paradise, be established within each of the national parks. These would be available for sale as plumes or live birds, on both the local and overseas market. This would be a better policy to ensure the preservation of the species than the present rigid (but unpoliced) controls. Similar policies could be adopted for birdwing butterflies. Host plants *(Aristolochia spp.)* would be established in large numbers within the parks and these would greatly increase the populations of any threatened species. In the absence of a large, well staffed and equipped zoological gardens in P.N.G., this becomes even more essential.

Recreational and educational parks could perform a second major function. It is suggested that a systematic effort to establish collections of all New Guinea plants in educational parks should be made. Careful labelling of collecting localities would be essential . It would be important to establish collections of the species from as many localities as possible in order to establish viable gene pools. Distribution of live material to overseas botanical gardens would greatly assist in both the study and preservation of New Guinea species.

Such recreational/educational parks could establish a certain financial viability if the proposed breeding establishments and plant nurseries were established. The parks should be able to sell native birds (e.g. Cassawary bred in W. District parks sent to Highlands for sale), animals, insects and plants, both for export and internally. This money would be channelled into a fund for the establishment and further development of park sites.

The areas of recreational/educational parks should be at least 1000 to 2000 Ha. Some 300 to 400 Ha could be devoted to the economic production of plants, animals, butterflies, birds and insects for sale. These areas would be open to the public for examination, basically for the resultant educational advantages. These areas would have little value for habitat conservation (except for restocking new areas) but would play a major role in species conservation. The remainder of the park would be left in its 'natural' state to give the people 'wilderness experience'.

2. SERAL RESERVES

Seral reserves will require quite distinctive management to either the first or the third category.

The widespread occurrence of secondary forest and grassland in Papua New Guinea has been discussed. If representative vegetation types are not regularly burnt or selectively logged to simulate garden activity, then many secondary habitats will disappear as a result of plant succession. If areas, e.g. on Mt. Wilhelm, are not regularly burnt, then the regeneration of the shrubs and grasslands will eventually result in considerable decrease in the length of forest margins and the high elevation flora which is restricted to the grasslands about the forest margin. Low frequency of important food plants will give related decreases in the bird and animal populations. Regular firing and controlled selective logging (to create canopy gaps) should be an integral part in the management of these parks.

Conservation of several forest types *(Eucalyptus deglupta, Castanopsis, Anisoptera)* will require catastrophic logging to preserve the ecological types.

3. NATURE RESERVES

Nature reserves will be required to preserve areas of climax forest. These should be disturbed as little as possible and be open for the public only in restricted areas. Scientific research on habitat types should be actively encouraged in parks within the last two categories. Specific areas would be reserved to preserve selected plant or animal species. The species should also be introduced to recreational or educational reserves for breeding and wider distribution.

AN EXAMPLE FOR SELECTION OF SITES FOR NATIONAL PARKS

It is proposed that sites be selected for National Parks in the vicinity of all major towns throughout Papua New Guinea.

Using Lae as an example, it should be possible to purchase four distinctive sites each worthy of consideration as a National Park.

1. Labu and eastern slopes of Herzog Ranges, possibly to Mt. Missim.

2. A coastal and hill forest area on the southern coast of the Huon Peninsula between Lae and Finschaffen.

3. A maritime park to include the islands off the Morobe coast - Juwani, Lasanga, Weleawa, Fly Islands, etc.

4. Lake Manum - lowland grassland types.

Class 2. would be the main educational/recreational park for Lae. Regular slide shows, guided tours, breeding of animals, birds, etc., plant propogation, should be established. The four proposed areas would retain a good sample of the lowland forest communities.

Areas for higher altitude communities - *Araucaria* - Bulolo, Lake Triste, Mt. Amungwiwa: *Agathis alba* stands - summit regions of Mt. Salawaket and part of the north slopes of the Salawaket Range under Podocarps and *Nothofagus*.

Habitat studies should be made in all national parks - vegetation plots established in detail and corresponding studies made of the bird, animal and insect floras. Such studies have yet to be conducted in Papua New Guinea.

CONCLUSION

It is essential in any approach to habitat conservation, to plan park management. The following must be taken into consideration:

1. The extensive areas of secondary vegetation types in P.N.G. requiring quite different treatment than that normally applied in parks.

2. The recognition of both the educational and scientific value of national parks.

3. The necessity to establish a viable economic base for the parks. This will also provide employment opportunities.

Catastrophic events are an integral component of New Guinea vegetation, these must be sustained in selected areas to preserve the present structure, physiognomy, and floristics of the communities.

REFERENCES

White, K.J. (1975). The effect of natural phenomena on the forest environment. Presidential address P.N.G. Scientific Society.

Johns, R.J. (1972) "Vegetation" in Encyclopaedia of Papua New Guinea.

Specht, R.L., E.M. Roe and V.H. Boughton, (1974) 'Conservation of Major Plant Communities in Australia and Papua New Guinea' Aust. J. Bot. Supp. Series No. 7.

ECOLOGY AND CONSERVATION : THE REGIONAL DIMENSION

Arthur Lyon Dahl
Regional Ecological Adviser
South Pacific Commission
Noumea, New Caledonia

Interest in the environment and conservation has been developing rapidly in the South Pacific over the last few years. The Regional Symposium on Conservation of Nature, Reefs and Lagoons, co-sponsored by the South Pacific Commission (SPC) and the International Union for the Conservation of Natural Resources (IUCN) in Noumea, in 1971, brought to a focus the needs in this area. The Thirteenth South Pacific Conference in 1973 approved a Special Project on Conservation of Nature within the SPC Work Programme, as part of which a Regional Ecological Adviser was recruited by the SPC in mid-1974.

Since then, nearly every Pacific Island country and territory has taken steps towards improved environmental protection and conservation. A terrestrial biological inventory is underway in American Samoa; the Cook Islands have adopted Conservation legislation, and offered an island as a world marine park; Fiji is to appoint a Government Ecological Adviser; French Polynesia has established major reserves on Tahiti; the Gilbert Islands have adopted new wildlife legislation; Niue has developed a comprehensive environmental protection programme; the Solomon Islands have requested assistance with conservation and environmental protection legislation; the Trust Territory of the Pacific Islands has begun inventories of historic and natural sites; and Western Samoa has adopted National Parks legislation and had a comprehensive survey for National Parks and Reserves. In many respects, Papua New Guinea has led the way with major national park, wildlife, and environmental programmes, as is so evident in the other papers being presented here.

The South Pacific Conference on National Parks and Reserves, hosted by the New Zealand Government in Wellington, in February 1975, with the co-sponsorship of SPC and IUCN, provided an opportunity for high-level discussions among Pacific Island countries and territories, further stimulating conservation activities.

Meanwhile, the South Pacific Commission has been developing its Environment Programme to encourage and support these advances to many of which the Commission has directly contributed. It is working to increase research in the area on major biological resources such as coral reefs and mangroves, as well as assisting with co-ordination and information exchange between existing research programmes. It is developing techniques for the monitoring of island natural resources, so that governments will eventually have reliable data on the status of their environment. It is surveying training needs in the region to see where new or supplementary programmes or courses may be needed. The Regional Ecological Adviser has consulted at length with almost all island governments on the organization and co-ordination of their environmental activities, and has prepared draft legislation for consideration by several governments. The need for ecological inputs into development planning, and the evaluation of environmental impacts, have been emphasized. The SPC is now preparing a Regional Ecosystem Survey as a framework for regional conservation planning, and has assisted in developing park and reserve proposals. It has particularly encouraged the study of traditional environmental management practices in island cultures, which can provide useful models and much data for the future. It is planning a programme of curriculum development assistance for environmental education suited to the needs of island children at the village level. It has assisted in the development of a museum programme in environmental education. It has explored the need for a regional scientific and technical information system, and begun publication of an Environment Newsletter as a medium of exchange between the many individuals and programmes active in the Pacific area.

In addition to its own programme, the SPC is co-operating actively with IUCN and the United Nations Environment Programme (UNEP) to increase the flow of assistance to the region. IUCN has funded the Regional Ecosystem Survey, and will assist with the Second Regional Symposium on Conservation of Nature being organized by the SPC in Apia, Western Samoa, in June 1976. UNEP has expressed great interest in the Pacific Island area and can be expected to provide increasing help in the future.

It is hoped that this regional dimension to activities for environmental management and conservation will help to increase the effectiveness and scope of national and territorial programmes, and will lead eventually to the development of island societies able to meet the needs of their people while maintaining a harmonious balance with their distinctive natural environments.

NATIONAL PARKS & RESERVES.
PAPUA NEW GUINEA'S NEEDS

by

M. A. Hill
National Parks Service

In his recent budget speech Mr. Chan said that management of natural resources for maximum national benefit was one area where the Government could not afford to make mistakes. This country's potential wealth is tied up in its natural resources of water, timber, minerals, fish, wildlife etc. In order to generate the necessary cash flow these resources will have to be exploited.

As Papua New Guinea strives to attain economic independence more and more exploitative processes, both large and small, will be established. All, to a greater or lesser degree, have the potential to degrade the environment. It is therefore essential that a sound environmental approach be taken to these "developments". Only if this approach is taken can we be sure of holding in check or minimising the potential adverse environmental effects.

A sound environmental approach to economic development is important in all countries. How much more important is it for Papua New Guinea whose lack of major development in the past has placed it in a unique position to learn from the mistakes of other countries.

There is no acceptable substitute for sound overall environmental management. A part of this management and an essential part is the setting aside of areas of the "natural" environment as national parks or reserves. National Parks are by no means a panacea for environmental ills, even though this sentiment seems to be expressed by some conservationists in other countries. However they are extremely important in the overall scheme of managing the environment to provide the optimum conditions for mankind's existence on this planet.

In many countries of the world national parks and reserves have been established as a last ditch resort to combat widespread environmental degradation. They have often been established over the only areas left even relatively untouched by technological man and frequently over areas that no one else requires for economic purposes. A poor way to establish a system of parks and reserves but better than having no system at all - especially for those countries that have no options in the matter.

Papua New Guinea still has many options open to it. It still has very extensive areas of natural environment from the coral reefs and atolls to the high alpine tracks. It is important for the future well being of this country and its people that, as part of an overall environmental management plan, extensive areas are set aside as reserves.

Throughout the world today approximately 145 countries have accepted the national park concept and modified or adapted it to suit their particular conditions. It is generally agreed that the main benefits that accrue to these countries from the establishment of a system of parks and reserves include the following:

1. National parks and reserves provide attractive natural settings within which visitors, foreign and domestic, can appreciate local scenery, vegetation, wildlife, history and culture.

2. They help provide for the growing outdoor recreation needs of increasingly urbanised populations.

3. They provide scientific reference areas on which adequate monitoring of changes due to agricultural or forestry monocultures, industrial development and other large scale environmental disturbances can be based and other scientific research carried out.

4. They provide gene pools - the widest possible range of types of plants, animals and other organisms any of which may be essential to the future well being not only of local populations but of all peoples, through their genetic potential.

5. They provide future generations with the opportunity to make land use decisions unaffected by previous utilisation of the areas.

6. They provide a bridge with the cultural post so that peoples undergoing very rapid and often unasked for changes in their lifestyles have areas where they can restore themselves in an accustomed and stable environment.

7. They provide ideal areas for environmental education, at all levels.

These benefits are certainly relevant to the Papua New Guinea situation. A system of parks and reserves could provide all those benefits listed above and many others too. However to achieve such a satisfactory state of affairs there are, as I see it, ten contexts in which the establishment of parks and reserves should be given consideration. Many of these overlap but all are sufficiently important to warrant separate mention.

The first is the most obvious and easily argued and concern the setting aside of areas of outstanding or even unique natural features - both physical and biological. Few people would query the desirability of protecting a country's unique natural features, in theory anyway, but the practical attainment of the protected status is not necessarily easily come by. However once established it is possible for all the benefits listed above to accrue from the setting aside of these areas.

Secondly there is the need to set aside areas containing representative samples of the natural features of the country. It is a relatively simple matter to generate interest in the truly outstanding or unique but much less simple to generate interest in the mundane and unspectacular. However it is of the utmost importance that representative samples of all the various habitats be set aside within a system of parks and reserves. It is in fact, essential if the conservation of gene pools with the widest possible range of types of plants, animals and other organisms is to be realised since the outstanding and unique are hardly likely to be representative.

Thirdly there is the requirement to set aside areas of our marine environment. Obviously marine parks and reserves will be covered by the first two categories. However I feel it is important to stress the need to give special consideration to these areas. Because of familiarity with the terms national park and reserves used in connection with the terrestrial sphere the marine sphere often tends to be over looked. Hence the need for special emphasis.

It is impossible to place too much stress on the need to set aside marine and coastal parks and reserves, particularly so far as Papua New Guinea and its neighbours in the South Pacific are concerned. Marine resources are of enormous significance to these countries and therefore the need for sound conservation of these resources, which must include setting aside designated areas, cannot be overlooked.

Considerable legislative and management problems arise with marine parks but these must be overcome and we should not permit the problems likely to be met deter us from pursuing this goal. The possibility of regional parks should certainly be investigated in the marine sphere.

By naming marine parks separately I do not want to give the idea that their management etc is quite separate from their terrestrial cousins. The opposite is the case as the marine park is vitally affected by activities on land the two cannot be divorced. The interface between the sea and the land does not separate the two elements but rather joins them together.

Next on our list is the need to set aside areas for the purpose of species preservation. In this context I refer to species of flora or fauna. We all know that the easiest way to exterminate a species is by destroying its habitat. The best way of ensuring that species threatened with extinction survive is by providing areas of suitable habitat sufficiently large to enable the establishment and maintenance of self sustaining communities.

The fifth context proposed for consideration is the setting aside of reserves in conjunction with major development projects be they agricultural, forestry, mining or industrial.

It is of particular importance where large scale agricultural or forestry development takes place because of their extensive nature. Such areas should be sufficiently large to be ecologically viable even while land use practices going on all round them are quite incompatible.

Next we should consider the need to set aside areas close to the main centres of population, throughout the country. All major towns and cities should have areas of natural bushland available for the outdoor recreation needs of their citizens. Such areas could also play a significant role in environmental education programmes because of their proximity to schools etc.

Because of the relative newness of urban living in this country the outdoor recreation needs of town dwelling Papua New Guineans are not large at present. However these needs will grow and it is important that suitable areas are obtained now while they are still available.

The next two contexts are very closely related but in my view require separate mention. These are, firstly, the setting aside of areas of cultural and pre historical significance and, secondly, the setting aside of areas of historical significance. No particular date can be set for the division of historical and pre historical as it would vary in different parts of the country. However history would cover the period since European influence and settlement and pre history all that period prior to such settlement.

The need for conservation of the culture of Papua New Guinea is recognised by all. This can be greatly assisted by the protection of sites or areas of great cultural significance. Sites of great pre historic interest are obviously an essential element of any programme.

An argument, however, can be made out against the need for protection of sites or areas of historical interest. Whether one approves or disapproves of the 'colonial' era it is an integral part of the history of this country and should be viewed as such. We must, therefore, give consideration to setting aside areas in this context.

Papua New Guinea is criss-crossed by an intricate network of walking tracks, many of great significance. Many of them were the main trade routes for the flow of goods from the coast to the Highlands and vice versa. In addition, of course, they were the main roads of the country in pre-European contact time. Consideration must be given to perpetuating these routes and walking tracks. Their disintegration and disappearance, over a period of time, would be both certain and tragic if nothing was done to set them aside. This context could well have been included with that concerning cultural and pre historical areas but I feel it deserves a category of its own.

Finally I come to the last context. This is concerned with the need to set aside natural areas in a "Land Bank". The idea that national parks and reserves are inviolate and "for ever" is widely held in many of the developed countries. Such an idea is in reality sheer delusion. National parks and reserves are only for as long as the pressures for their alternative use remain below a certain level. Once that level is reached politicians have to act and, all at once, your park becomes a housing estate or a mine. It is, however, perfectly right that the parliament of the country should be able to revoke the status of reserves in the national interest. It is important, though, that the process to achieve revocation be made as difficult as possible.

It is not difficult to foresee that future generations are likely to inherit an environment that has been well and truly "worked over" by previous generations in their exploitation of the country's natural resources. The only areas of "natural" environment will be in parks or reserves or areas that have no economic use because of extreme ruggedness or inaccessibility. The land use options available to these future generations will therefore be minimal.

Consideration must be given now to the question of putting natural areas into storage for the benefit of future generations whose land use options may there by be maximised. If such a course of action was taken now the parks and reserves we establish would have a very much longer life expectancy. Probably a utopian vision but nonetheless worthy of consideration for all that.

It is easy to write down on paper the requirements of Papua New Guinea in the establishment of a system of parks and reserves. It is not so easy to fulfil those requirements, in reality. Many problems are encountered in this country which do not occur in other countries that have accepted the national park concept and the need to establish reserves. As has occurred in other countries it is necessary to modify and adapt the concept to suit the conditions existing in Papua New Guinea. It is most important that a flexible approach be taken and that we do not get bogged down by terminology. A national park in Papua New Guinea is not going to be the same as a national park in New Zealand or Canada for instance, but it will be none the less a national park for that.

Particularly in areas with low population densities Papua New Guineans lived, and, in some areas, still live as part of the ecosystem and not outside it as is the case with people of more advanced technology. Therefore where this situation coincides with a proposal to establish a national park there is no reason to alter the people's way of life. They should be permitted to live within the park or reserve (if such is established) for as long as they wish and so long as their way of life remains a part of the ecosystem.

This is just one example of the need for flexible thinking and the need for those concerned with conservation not to be hidebound by convention or conventional terminology. It is most likely that to achieve a system of parks and reserves worthy of this country a different approach may have to be taken in each case. With 97% of the land area in this country under customary ownership the potential for establishing parks and reserves on government owned land is very small. Over a period of time all the major parks and reserves will be established over customary land. Different owners will have different aspirations and requirements. Some may be willing to give up all their rights to the land and sell out right to the government for conservation purposes, thus retaining basic ownership. Some may be prepared to lease land to the government on the condition they can retain certain traditional rights to the land but enter into an agreement with the government as to how the land should be used. In other situations reserves may be established by dedication or the establishment of a conservation easement without alienation of the land and so on.

What we are really after in Papua New Guinea is a sound environmental management programme incorporated into which is the establishment of a system of parks and reserves. It is probable that, to achieve this, each one of the methods mentioned above, and others not mentioned, also will be used in the varying circumstances encountered. It is imperative that this flexible approach be maintained and that legislation to permit such a milti-faceted approach be enacted.

The other great need in order that a system of parks and reserves can be established is that of public awareness and support. Ultimately any conservation programme will stand or fall by the support or lack of it from the people.

In effect good management of national parks is very difficult even under the best conditions. Good management of a national park established in an area where the local people are resentful or hostile is virtually impossible. In the wider context of a nation, the sound management of the environment can only be achieved with the support of the populace. It is not our duty to force conservation practices on the people; rather it is our duty to ensure that people are fully aware of the environmental problems posed by development and the ways and means of mitigating against them. They should also be made aware of the land and resource use options available to them. The best conservation schemes evolve from the bottom up and are not imposed from the top down.

I have tried to briefly set out the needs, as I see them, of Papua New Guinea for the establishment of a system of parks and reserves as an integral part of a conservation programme for the sound management of this country's environment. Before we can even begin to meet these needs, there is one essential prerequisite that must be met. This is the commitment of the government to such a policy. Any conservation programme must be implemented within the guidelines of government aims and policies. Therefore initial and continuing government support for the principles of conservation is the lynch pin of any programme for the sound management of our environment. In today's world, so much more aware of the irreparable harm done to the environment through unplanned or thoughtless utilisation of its natural resources, this support will almost certainly be forthcoming. After all we are, everyone of us, concerned that this and future generations have the opportunity to live out their lives in the most favourable conditions obtainable.

NATURAL AREAS (ECOLOGICAL AREAS) IN THE PACIFIC

by

J. L. Gressitt
Wau Ecology Institute

ABSTRACT:

Natural or ecological areas serve a variety of aims and supplement national parks. Mainly they are to preserve samples of ecosystems and do not primarily serve purposes such as recreation, camping, etc. They often tend to be devoted to base-line studies, monitoring of changes, and to teaching, as well as preservation of ecosystems and species.

Natural area systems (ecological areas) have been gaining much popularity in recent years. Many countries, including various ones in the Pacific, have systems established and are constantly adding new areas to their systems.

The concept of natural areas, or natural area reserves, ecological reserves or nature sanctuaries, is one that complements a national park system. Whereas a national park may comprise an area with a combination of several special attributes, it does not necessarily include all the ecosystems of the general area. Moreover, the representative areas for some or most of the ecosystems in a national park may become overrun and modified as a result of the activities of large numbers of visitors, and various component species may disappear. Also in a national park some animals may greatly increase from protection with reduced predation, and may ravage certain types of plants, and upset the ecosystem.

A natural area may be selected to protect an ample area of a single ecosystem, or it may comprise more than one. In any case the area involved per ecosystem must be adequate for safe perpetuation.

Natural areas may have one of several quite different purposes, or may combine more than one purpose. The various purposes and types of natural areas may be outlined as follows:

(a) To preserve a representative sample of a given ecosystem in a manner to give it permanent protection. Size must be adequate to prevent losses of species, giving consideration to those species which range widely for their food or breeding. Often such an area may be contiguous with a national park or national forest, or may be within one or the other, but may be restricted from public entry. It should be protected from hunters and may preferably have some natural protection as well as laws, patrolling, or fencing as necessary. In cases where feral animals are damaging the environment, hunting of these should be permitted or encouraged.

(b) To protect areas containing endangered, rare or local species. Such areas may not necessarily represent typical or relatively untouched ecosystems, and in some cases the endangered species may still be surviving because of human modification of environment, rather than in spite of it (as with *Ginkgo* and *Metasequoia* in China). Thus this type of area may be termed a gene pool reservoir, a genetic bank or a natural live museum. The latter term may also be applied to category (a).

(c) To protect areas for research. In general these are relatively unmodified areas to serve as the base-lines for long-term comparison with adjacent or similar areas that may be subjected to various degrees of human influence. Besides permitting the basic research which will determine the characteristics and requirements of the environment, it will enable comparisons with equivalent areas undergoing modification, and will assist in determining conservation measures.

(d) To protect areas which are samples of ecosystems already modified by man, to study their possible recovery and reversion to natural circumstances. This category may also be applied to areas affected by natural phenomena or disasters, such as volcanic activity, tsunamis, floods and landslides. Thus this category and the preceding represent "outdoor laboratories."

(e) To protect areas for teaching purposes. These may duplicate some of the above categories and thus be to some extent expendable, but should be maintained in as natural a form as possible in order to serve as continuing examples for structure, make-up and functions of ecosystems, or for their management and rejuvenation. Thus these are "outdoor classrooms."

(f) To protect areas demonstrating some combination of unique botanical, zoological or geological phenomena. This might include a beach, reef, islet, waterfall, lake, volcano or lava flow with adjacent areas having special local associations or natural and modified environments side-by-side.

Some natural areas may combine some of the above purposes or might be divided in equivalent portions, part to preserve the ecosystem untouched, part to research it, part to use for teaching, and a part to monitor change or to test management practices.

As an example of category (f), one natural area in Hawaii includes the entire extent of the last lava flow on Maui, which flowed into the ocean, forming two bays. The area thus includes the lava flow, pioneer vegetation on parts of it, ponds and caves of various salinities, with unique biota, and a very rich inshore marine area where scuba diving (but not killing) permits observing a great variety of marine life in action.

Natural areas may often be smaller than national parks; and some may be relatively inaccessible to the public. Extra large areas may be called Wilderness Areas, and may preserve several ecosystems. In general, roads, buildings and gardens are not permitted within areas, so people may enter only on foot, and must observe rules. Other types of parks will cater more to recreation, camping and hiking.

Papua New Guinea needs a variety of natural area reserves in order to protect adequate samples of all its ecosystems. By protecting these, all the component species should be preserved. It must be stressed that the Birds of Paradise cannot be protected by legislation alone, but require their habitats to be preserved. One of the purposes of natural reserves is to restrict indiscriminate or wide-scale use of undisturbed land for exploitation before all ecological consequences have been considered. Natural areas are to be dedicated permanently to conservation, and this designation must be made very difficult to modify. Otherwise there will be no real conservation. Natural areas need to be documented, integrated into a planned system, and adapted to local circumstances and requirements. Some of these natural areas may be set up as provincial or local parks.

A general trend in the development of natural area reserve systems is for communities to make suggestions for areas for protection, for natural history societies and others to search out such areas, and for conservation committees to survey and plan them. Often, public (non-governmental) conservation organizations raise money to buy the land. Some states effect reserves almost independent from government, with government approval. In other cases, all matters are passed by a state conservation board, or committees with representatives from relevant government departments, like forestry, wildlife, conservation, lands, etc. An important thing is that such committees and boards must include scientists or others with sufficient understanding of the ecological principles on which conservation must be based. Agreement of the local population must be obtained by public hearing coordinated with government, with the local people sufficiently informed of the value and advantages of such a program.

Wau Ecology Institute would like to suggest natural areas at Mt. Kaindi, Mt. Missim, Bulldog Road with the Ekuti Range (Mt. Amungwiwa), and Morobe (marine, reefs, isles, shore, estuaries, rivers, swamps and lowland forest). Also comparable alpine, mid-montane, lowland and marine areas in the Western, W. Sepik, Northern, Gulf, Milne Bay, and the Island provinces.

It has been stated that Mt. Kaindi is too disturbed for saving. I can graphically refute this by example of a still smaller mountain I have been visiting over a period of 46 years. It is the nearest mountain to the largest city in the world: Mt. Takao, one hour by express from downtown Tokyo. When I started visiting Takao it was in the country, an hour by train from the edge of the city, with no highway. Now there are two train lines, paved highway and both a cable railway and cable lift go nearly to the summit, besides four walking tracks. Some 20-storey apartment buildings are close to the base of the mountain. 3,000 or more people were on the mountain the day I climbed four months ago, but the mountain looked better than it did on my preceding visit 10 years earlier: less human waste and debris around, though now many signs, refuse pits and toilets. The entire mountain is forested. Part consists of tall forests of old trees, and part is cyclically harvested (*Cryptomera* poles), but with methods leaving one with quite a reasonable impression of complete forest cover. Some of the refuse pits are built on the slopes below trails from flattened drink tins, stacked and camoflaged with dirt or bark. No collecting of specimens is allowed, and research is by special permit. There is a zoo near the top, and educational signs with paintings of the birds and mammals to be seen in the wild, and names on trees along the tracks. It is a very great asset for field outings for schools, as well as for educating the general public re: fauna, flora, ecology and conservation. An excellent museum is at the foot of the mountain. This and the signs, etc are sponsored by a local natural history society.

Many countries in the Pacific area now have or are developing natural area systems. Perhaps the most developed of these is that of British Columbia, with now 62 of a projected 100 reserves established. These vary from 0.7 to 11,000 hectares in size, and are of 5 types:

(a) Scientific and educational;

(b) representative of natural ecosystems;

(c) examples of ecosystems modified by man - to study recovery;

(d) areas with endangered species;

(e) areas with unique botanical, zoological or geological phenomena.

To conclude, I quote Resolution X of the 13th Pacific Sciences Congress, 1975:

"WHEREAS the Pacific region possesses distinctive flora and fauna including numerous endangered species and a broad range of ecosystems;
WHEREAS animal and plant life forms, both terrestrial and aquatic, are biological riches that are vulnerable to the growth of population and technology, especially on isolated islands;
WHEREAS these unique natural resources must be preserved, both to provide baselines against which environmental changes can be measured, and for the enrichment of future generations of mankind;
WHEREAS systems of ecological reserves can support natural communities large enough to remain as unmodified samples of each country's special living heritage; and
WHEREAS special programs on Pacific ecosystems and ecological reserves have shown that, while a few governments, such as that of the Canadian Province of British Columbia, have a well advanced network of reserves, many countries are in the earliest stages or have not yet begun;

"THEREFORE, be it resolved that all countries be urged to place the highest priority on planning, designing, and effectively protecting a system of ecological reserves."

POTENTIAL PRODUCTIVITY OF CORAL REEF FISHERIES

by

J. L. Munro,
Biology Department,
U.P.N.G.

Efforts towards the development of the fishing industry of Papua New Guinea have in the past been directed almost exclusively towards development of the fisheries for tuna, spiny lobster (crawfish), shrimp and barramundi. All of the these fisheries have offered immediate possibilities for the export of high-priced marine products and consequently, almost all efforts have been (and continue to be) directed towards their development, improvement and long-term conservation.

These developments have tended to obscure the importance of the generalised coral reef fisheries as a source of protein. The point has now been reached where there is an urgent need for inexpensive proteins in many inland areas of Papua New Guinea but the possibility of harvesting reef fishes at any level substantially above the subsistence level appears to not have been seriously pursued up to the present time.

The lack of attention paid to coral reef fisheries stems from two basic factors:

1. The popular misconception that whilst coral reefs might harbour a bewildering diversity of life, the harvestable component of the reef fauna is extremely limited and is rapidly exhausted by any intensive exploitation.

2. The very real difficulties of marketing and processing an assortment of fishes which differ substantially in shape, size, texture, and palatibility and, in their suitability for processing.

The purpose of this paper is to show, on the basis of recently acquired data, that the potential harvest from coral reef areas is greatly in excess of that which is commonly believed to be possible, and to indicate several means whereby better utilisation of the reef fish resources might be achieved.

To put the following arguments in their true perspective a few comparative facts will be useful.

Stevenson & Marshall (1974) reviewed a series of estimates of fish production from coralline areas and found estimated annual harvests to range - between 4 kg/ha and 50 kg/ha and suggested that average harvests in the more fully developed reef fisheries might normally range between 20 and 50 kg/ha and that even moderate exploitation would normally yield about 10 kg/ha. They cite as examples of high productivity the harvests taken in Mauritius (47 kg/ha/yr) and at Ifaluk atoll (51.4 kg/ha/yr). Less densely populated areas or islands with extensive shelf areas, or sophisticated consumer economies commonly achieve 8 - 20 kg/ha of shelf, even when only a select proportion of the catch is considered to be marketable. Examples include

TABLE 1.

Estimates of potential annual harvests of marine animals from shelf areas bordering maritime provinces of Papua New Guinea.

			Potential (metric	yield tons)
	sq.nautical miles	thousands of hectares	Moderate exploit.	Heavy exploit.
Western Province - Torres straits area*	9,850	3,378	27,000	67,660
Western & Gulf Prov. -Gulf of Papua **	13,350	4,578	36,600	91,500
Central Province	1,700	582	4,660	23,300
Milne Bay Province	12,000	4,115	32,900	93,300
Morobe Province	720	246	1,970	7,390
Madang Province	260	89	710	2,670
East Sepik Province	185	63	510	1,900
West Sepik Province	150	51	410	1,540
Manus incl. Hermit & Ninigo Islands	2,150	737	5,900	22,100
New Ireland Inc. Lyra Reefs (500 nm^2)	1,300	446	3,570	13,400
West New Britain Province	1,280	439	3,510	13,200
East New Britain Province	630	216	1,730	6,490
Bougainville Province	2,240	768	6,140	23,000
TOTAL	50,665	17,370	138,910	388,490

* Bounded by longitude 141°E in the west, latitude 10°S in the south and by a line joining Daru, Bramble Cay, Anchor Cay, and Murray Island in the north and east.

** Bounded by the abovementioned Daru-Murray Island line in the south-west, by the 200 m. isobath in the south and the Gulf-Central District border in the east.

Jamaica, Cuba, Guadeloupe, Martinique, St Lucia and Barbados in the Caribbean and numerous islands in the Pacific (Stevenson & Marshall, 1974; Munro, 1973).

Munro & Thompson (1973) have used a modified exponential surplus yield model to demonstrate that harvests from various sectors of the coralline Jamaican shelf increase with increasing fishing intensity to a maximum of 43.2 kg/ha/yr (at a fishing intensity of 3.09 canoes/100 hectares) but that yields decline at higher fishing intensities. This yield is largely achieved by fishing with small-meshed (4.13 cm) fish traps. It has recently been shown that this yield could, in theory, be substantially increased by appropriate management techniques, including an increase in the mesh size of the traps (Munro, 1975).

As pointed out by Stevenson & Marshall (1973), yields of this magnitude (40-50 kg/ha) are equal to or in excess of those normally attained on the great industrial fishing grounds of the North Atlantic ocean, including such areas as the Icelandic banks, the North Sea and Georges Bank.

Current estimates of annual catch by the subsistence and artisanal fisheries of Papua New Guinea amount to about 13,600 metric tons (30 million lb.) (D.A.S.F. mimeo report, 1975). The basis for this estimate is unknown, but as the total shelf area of Papua New Guinea amounts to over 17 million hectares (Table 1) the current yield is probably less than 0.8 kg/ha/yr.

Table 1 shows details of the estimated areas of shelf or adjacent oceanic reefs which border the various provinces of Papua New Guinea (the Torres Straits area and Gulf of Papua are treated as separate entities). As can be seen from the table the relative extent of shelf is highly variable. In addition, the extent of coral reef development is not uniform. The description given by Whitehouse (1973) indicates that the Torres Straits area is generally coralline but the reefs appear to not be prolific. The Gulf of Papua is devoid of coral growth owing to the high turbidity and the great freshwater inflows and only the Central, Milne Bay, Bougainville, New Ireland and Manus Provinces harbour extensive reefs. The Northern Province has a wide shelf but relatively poorly developed reefs and New Britain, Morobe, Madang and East and West Sepik Provinces are characterised by extremely narrow shelves with discontinuous reef development. The reasons for this pattern of development are not clear but appear to be related to tectonic factors and the steepness and geologically unstable nature of the shorelines (Whitehouse, 1973).

On the basis of the present evidence it is therefore possible to make some generalised estimates of the likely extent of reef fish resources in Papua New Guinean waters. It is highly likely that all of the areas listed in Table 1 could produce a minimum of 8 kg/ha if subjected to quite moderate levels of exploitation. For the areas richly endowed with entensive reefs, much greater potential productivity is possible given the necessary market outlets for the harvest. It seems likely that given rational management there should be no obstacle to eventual attainment of a production figure of 40 kg/ha in the Central Province. It is likely that such production would arise mostly from the activities of artisanal fishermen operating from small craft in the general vicinity of their home bases and supplying fish transporters which would transport catches to Port Moresby or other markets either by road or by water.

In contrast, development of the fisheries in the very extensive shelf areas of the Milne Bay and Northern Provinces would have to rely upon larger vessels operating in areas remote from their land bases, and remote from major markets. Such fisheries require greater operating and distribution costs and it is therefore inevitable that exploitation will be at a much less intensive level. Nevertheless ultimate yields of 20 kg/ha/yr appear feasible. A similar yield is likely for the non-coralline Gulf of Papua (Gulland, 1971). It seems likely that all of the other provinces could also produce harvests of about 30 kg/ha/yr given the necessary marketing structure and incentives to development.

The projected maxima for the various regions which have been outlined above will only be attained at such time as markets for the products develop or population increase in the area increases demand. Clearly, the major potential local markets are in the New Guinea Highlands and the main urban centres. It seems likely that an extension of the prawn trawling industry in the Gulf of Papua to include harvesting of fishes would rapidly saturate the Central Province markets and that the absence of road communications with the Highlands would preclude economical distribution of inexpensive processed fish products to the Highlands. In contrast, the Milne Bay and Norther Provinces are strategically situated to supply the Highlands markets via land routes starting at Lae. The critical area here relates to the production of an acceptable cheap processed product which can compete with canned mackerel from Japan and pilchards from South Africa. Either a dried salted product (whole fish, fish sausage or fish cakes) or an inexpensive can of "mixed fish" would appear to be indicated. Finally, the possibilities of exporting high quality frozen fish or other processed fish to South-east Asia should not be overlooked. The constant presence of Taiwanese fishing vessels in Papua New Guinea's waters is probably indicative of the demand.

The main impediment to realising the potential harvest of marine products from the seas surrounding Papua New Guinea lies in the difficult marketing conditions, the low purchasing power of the major part of the population and the highly competitive nature of the present imports of canned Japanese mackerel and South African pilchards.

From the conservationist viewpoint it can be concluded that dermersal fish stocks in Papua New Guinea are virtually unexploited and in no danger whatever of over-exploitation at the present time. The present overall harvest rate is probably less than 4% of the potential.

REFERENCES

Gulland, J. A. (1971) The fish resources of the ocean. West Byfleet, Fishing News (Books) Ltd.

Munro, J. L. (1973) The biology, ecology, exploitation and management of Caribbean reef fishes. Part I. Coral reef fish and fisheries of the Caribbean Sea. Res. Rept. Zool. Dept. Univ. West Indies 3(1): 43 pp.

Munro, J. L. (1975) The biology, ecology, exploitation and management of Caribbean reef fishes. Part VI Assessment of the potential productivity of Jamaican fisheries. Res. Rept. Zool. Dept. Univ. West Indies 3(6): 56 pp.

Munro, J. L. & Thompson, R. (1973) The biology, ecology, exploitation and management of Caribbean reef fishes. Part II. The Jamaican fishing industry, the area investigated and the objectives and methodology of the ODA/UWI Fisheries Ecology Research Project. Res. Rept. Zool. Dept. Univ. West Indies 3(2): 44 pp.

Stevenson, D.K. & Marshall, N. (1974 Generalisations on the fisheries potential of coral reefs and adjacent shallow water environments. In Proceedings of the Second International Symposium on Coral Reefs Brisbane, Great Barrier Reef Committee 1: 147 - 156.

Whitehouse, F.W. (1973) Coral reefs of the New Guinea region. In Biology and Geology of Coral Reefs (ed. O.A. Jones & R. Endean) New York & London, Academic Press, 1: 169 - 186

TERRESTRIAL FOOD RESOURCES (PLANTS) OF PAPUA NEW GUINEA

By

T.N. Khan
Faculty of Agriculture, U.P.N.G.

Ecological variation in Papua New Guinea caused by the differences in soil, rainfall and temperature offers a scope to grow a variety of food plants (Table 1). However, in the absence of marked seasons as is the characteristic of Sub-Tropical and temperate climates, P.N.G. climate is limiting for a few useful plants which have longer photoperiodic requirement or which need extreme temperature differences between seasons for their physiological rhythm. Most of the food plants cultivated as staples have probably been introduced to this island.

Cassava, Sweet Potato, winged bean and Hyacinth bean are probably recent pre-European introductions. Others such as Taro, Banana, Sugarcane, Coconut may have been introduced in wild or cultivated form in antiquity and have now evolved a great deal of variation. Some authorities believe P.N.G. to be the centre of origin of Sugarcane and banana. Sago palm and Pandanus are semi-cultivated species and may be indigenous to P.N.G. A distinct feature of all the traditional crops in P.N.G. is the great wealth of genetic diversity.

A number of food plants have been introduced to Papua New Guinea since colonization and these include mainly three groups vegetables, cereals and legumes. By far the most important group of plants from the nutritional point of view is legume because of its high protein content. Cereals such as maize and sorghum although widely cultivated now, form negligible part of Papua New Guinean's diet. Rice consumption has increased without a parallel increase in the local production. Urban centres of P.N.G. are heavily dependent on imported foods including wheat, as now bread is a regular part of the diet. There is a possibility of growing wheat at higher altitudes above 7500' but terrain and level of technology is prohibitive. As most of the high altitude areas are relatively inaccessible, marketing may pose another problem. A man made cereal <u>Triticales</u> has been bred through a cross between wheat and rye, and combines the bread making quality of wheat with a greater range of adaptation. The above cereal has a distinct potential for high altitude valleys (1,200m - 1,500m) and humid lowlands.

The above discussion however does not include a variety of wild plants and cultivated green vegetables of indigenous origin (Powell *et al*, 1975) which form an important source of many dietary needs.

It is anticipated that with the growth in population a greater supply of local food will be needed. This will have to be met by both traditional and recently important crop plants. To meet this challenge it will be necessary to conserve the genetic variation in traditional food crops to provide a basis for selection and breeding work in future. If steps are not taken urgently most of this variation will be lost in favour of newly introduced genetically homogenous varieties.

****

REFERENCE

Powell, J.M. and S co-authors. 1975. *Agricultural Traditions of the Mount Hagen Area*. Department of Geography. Occasional Paper No. 12. University of Papua New Guinea, Port Moresby.

Table 1: Some Important food plants of Papua New Guinea. (Vegetables are not included in this list).

(A) TRADITIONAL

Tuber

Sweet potato	*Ipomoea batatas* (L.) Lam.
Taro	*Colocasia esculenta* (L.) Schott
Yam	*Dioscorea alata* L.
Cassava	*Manihot esculenta* Crantz

Legumes

Wing bean	*Psophocarpus tetragonolobus* (L.) DC
Kudzu	*Pueraria lobata* (Willd)) Ohwi
Hyacinth bean	*Lablab niger* Savi
Lima Beans	*Phaseolus Lunatus L.*

Miscellaneous

Banana	*Musa acuminata* Colla *M. Balbasiana* Colla Hybrid cultivated forms
Sugarcane	*Sachharum officinarum*
Coconut	*Cocos nucifera* L.
Sago palm	*Metroxylon sagus* Rottb. *M.* *M. rumphii* Mart.
Pandanus	*Pandanus jiulianettii* Mart. *P. brosimos*
Pitpit	*Sachharum edule* Hassk. *Setaria palmfolia* (Koen.) Stapf.

(B) RECENTLY INTRODUCED

Tuber

Potato	(*Solanum tuberosum* L.)

Cereals

Rice	*Oryza sativa* L.
Maize	*Zea mays* L.
Sorghum	*Sorghum bicolor* (L.) Moench

Legumes

Peanuts	*Arachis hypogaea* L.
Soya beans	*Glycine max* (L.) Merr
Cowpeas	*Vigna unguiculata* (L.) Walp.
Mung beans	*Vigna radiata* (L.) Wilczeck
Pigeon pea	*Cajanus cajan* (L.) Millsp.
Peas	*Pisum sativum* L.
Common Bean	*Phaseolus vulgaris* L.
Rice beans	*Vigna umbellata* (Thunb.) O. & O.
Black Gram	*Vigna mungo* (L.) Hepper

SOME THOUGHTS ON THE ECOLOGY AND CONSERVATION OF MAMMALS, REPTILES AND AMPHIBIANS IN PAPUA NEW GUINEA

By

J. I. Menzies,
Biology Department,
U.P.N.G.

INTRODUCTION

In the five minutes allotted, it is not really possible to say anything very specific about any of these animals so these notes simply reflect a few thoughts that I have on the subject.

It seems to me that there are two important points that make this country different from most others. The first is that this is probably the only large land mass that has no large native land animals. The second is that less is probably known about our native fauna than about that of most other countries. These two factors are obviously linked and must affect the way in which ecological studies and conservation measures will be tackled. Perhaps the only other country in the world with such a peculiar endemic fauna is Madagascar and the situation as regards many of its species seems rather desperate (Fitter, 1970).

THE NEED FOR RESEARCH

Large animals are relatively easy to study, especially when they live on open plains, and considerable progress has been made in other parts of the world by such institutions as the Nuffield Unit of Tropical Animal Ecology in East Africa (Beadle, 1974). Moreover, large animals have an international appeal and attract funds and zoologists which small animals do not. Apart from species that are now being studied by the wildlife section of D.A.S.F., few ecological studies have been made in this country.

The fauna of New Guinea has, however, attracted attention from early years and one might suppose that the taxonomy, at least, is in good order but in many cases, nothing is further from the truth. True, the vast collections made by the Archbold expeditions have resulted in some published monographs

(Tate on the rodents, dasurid and macropod marsupials for instance and Zweifel on microhylid frogs and McDowell on snakes) but Tate's work has limited application to field workers within the country and many other groups are badly in need of revision.

The fauna of New Guinea is largely a forest one and forest animals are harder to study than plains ones, especially when they live in trees and are nocturnal. It might be possible to compare our fauna with that of other countries equating, say, small primates with some of our marsupials but comparisons are limited. More validly, it might be possible to relate studies on Australian fauna to related species in PNG but there are generally such differences in habitat, even in closely related species that comparisons would probably need to be drawn with caution. For instance, a great deal is now known about the biology of the small echidna, *Tachyglossus* (Griffiths, 1968) but I don't know whether it would be valid to extend this information to the large species, *Zaglossus* which is already thought to be endangered and subject to special legislation. In many cases, the problem is undoubtedly lack of knowledge and not lack of animals - many of our species have an apparently patchy distribution and only a lot more research will produce reasons for that. *Zaglossus* seems to be common in some places - during the single week that I spent with the Forestry Department's Mt Suckling expedition in 1972, the camp attendants caught, cooked and ate at least eight *Zaglossus*. In some other places, in apparently suitable country, the animal is unknown and the lack of a vernacular name for it suggests that it never existed there, at least in historic times.

Research is urgently needed but the country does not have sufficient manpower resources to do it. The output of zoologists from the university is very small and many of our graduates are saddled with administration as well as research. Obviously, help must come from outside but it is not enough to rely on zoologists who come here to continue their own chosen programmes for their activities cannot easily be directed. They will come here to do what they want, or will not come at all. What is needed is some sort of endowment whereby fellowships can be offered (perhaps at post-doctoral level) to zoologists to come here and carry out specific research projects. We already have laboratories in various places, it is money and manpower that is needed.

ENVIRONMENTAL DETERIORATION

However much money is put into research, progress is bound to be slow - one cannot await the accumulation of data before proceeding with conservation measures otherwise, as in many countries, it will be too late.

Much of our fauna is a forest fauna and its status is obviously linked to the state of the country's forests. Forests and forest fauna is particularly vulnerable to human activity, and even in those parts of the country where the natural vegetation is not forest but woodland deterioration of the habitat can still go on. It is necessary to see what can be done in the meantime to prevent habitat deterioration as far as possible.

The general pattern of development nowadays appears to be a movement of people towards villages and towns rather than living in the scattered tiny communities previously favoured. Though this may lead to depopulation of the remoter areas and thus less pressure on land and fauna, the effect on the countryside around the settlements is dramatic. One has only to fly over the country to see the areas of deforestation around villages, patrol posts and so on. Obviously the move towards larger and large villages is going to continue but is all the environmental deterioration that accompanies it inevitable? Loss of tree cover is brought about by clearance for farming, firewood cutting and fire but I will confine my remarks to the second two. Over much of the country firewood is the only means of cooking, even in such large institutions as high schools it may be the preferred fuel, presumably because, in the short term, it is very cheap or even costs nothing. Wood is a renewable resource, there is nothing wrong in using it provided it is replaced but how many institutions using wood as fuel directly or indirectly encourage firewood plantations? High schools which ought (and perhaps do) include the principles of conservation in curricula happily send the pupils out into the bush to cut firewood without thought to the time when there would be none left. Natural regeneration may take place but, if clearing goes on in a place where forest is restricted to gullies, there seems to be little chance of natural replacement.

Fire is a subject with which those living in the Port Moresby area are increasingly concerned but the problem affects other parts of the country as well. Obviously the greatest effect is going to be in the drier areas where the natural succession can easily be diverted towards a fire climax of coarse grass and little else. All burnable areas of the world get burnt regularly, it is not a feature peculiar to Papua New Guinea but, because it is a traditional activity, I do not see why it should continue in a way of life that is no longer traditional. Reasons for burning are several - driving game is an obvious one or simply making it easier to walk from A to B in the dry season is another but much of the burning going on in the Port Moresby region is pointless and senseless so something ought to be done about it. I believe it is already an offense to burn grass near the town. I do not know how many people are prosecuted each year but the law is no deterrent for virtually the whole are suffers the annual ordeal by fire. Even if it is not possible to stop the burning we do not need to stand by and watch the country turned into a desert. Burning is rarely beneficial but uncontrolled burning is worst of all. The relatively harmless effects of early, controlled burning have been demonstrated elsewhere (Charles & Keay, 1960 or Hopkins, 1965). Has this been tried anywhere in Papua New Guinea? Another preventative measure might be the introduction of fire-resistant tree species. If we can't have hillsides covered with Eucalyptus woodland I would prefer to see them covered with introduced *Butyrospermum*, for instance, rather than just barren rocky outcrops.

When forest is converted into savanna by burning one might suppose that the native savanna spp take advantage of the additional habitat while the forest fauna regresses. This undoubtedly happens in the case of some lower vertebrates but, with mammals, I am not so sure. In many cases, I think it is exotic species that become established in the new areas and there are three small vertebrates which I think about in this context, all common in the Port Moresby area, elsewhere too, I am sure, but this is the area I am most familiar with. These are the black rat (*Rattus rattus*), the domestic cat and the cane toad, *Bufo marinus*. Schulze-Westrom (1970) deals at length with the problem of the toad but my concern would be more with the other two. I do not think we can do much about the toads, other countries have toad problems and I do not remember ever seeing any reports about successful control measures. Perhaps a research programme investigating some method of biological control would bring results but I would rather see the the money spent on direct research into some native species. I certainly do not approve of further introductions (the toad doesn't need much help anyway) but studies here suggest that the toad is more of a nuisance than a direct threat to native fauna. The highest toad population is around towns where they feed mostly on insects attracted to street and porch lights, which would die anyway.

Locally I think that black rats are a much more serious problem for they are certainly not confined to the vicinity of large towns. The commonest rat in the central district savannas (as deduced from accumulated owl pellets) used to be *R. sordidus*. Trapping here now yields only *R. rattus* (and *Mus musculus*). Direct competition? I don't know but native rodents have a very low fecundity (my own studies and Dwyer, 1975) and could quickly be overwhelmed by the prolific exotic rodents.

I do not know the extent of the cat problem either but the numbers of cats that one sees while driving around at night is high. Cats are like rats in that they are prolific but are unlike dogs in that they are perfectly capable of looking after themselves in the bush. I do not know what the feral cats around Port Moresby eat but I see no reason why they would avoid native rodents or marsupials, or ground nesting birds. Such animals, in other places, have been particularly vulnerable to introduced predators.

So what is needed is research, habitat protection and education, for without general support conservation measures will be ineffective. I do not include legislation at this point because I think that the existing necessity of obtaining permission to export any animal is adequate safeguard against depredations of the livestock trade. To legislate against internal predation is much more difficult (is prohibition on shooting of birds of paradise with guns effective?). Rather than have a law to protect, say, a rare lizard I would rather have one to prevent people turning out their unwanted cats. I do not think either would be easy to enforce.

Habitat protection

New Guinea's fauna consists largely of small animals which is an advantage because protected areas need also be small, whereas protection of large or migratory species (the turtles and dugong are probably our only species in this category) may require international cooperation. Large national parks are part of any country's constitution but are often difficult to establish because of the amount of land, or the ownership of the land involved. Large numbers of small reserves can be equally effective and are probably easier to establish. Moreover, the land does not need to be taken over by the central government, whose function should be technical management rather than outright ownership. It may not even be necessary to forbid hunting, at least initially, since this could generate opposition to reservation - though I would draw the line at nocturnal hunting from trucks fitted with searchlights, as practised around Port Moresby. The very fact of habitat protection means increased protection for the fauna, voluntary relinquishing of hunting rights can come later. The protected forest system of Sierra Leone seemed to me a good example of this sort of locally owned conservation. The individual "protected forests" were often quite small, merely roadside strips no more than 100 m or so wide. Such areas offered shade, windbreaks and sanctuary for fauna, and did not occupy much arable land. I suppose they could be considered as ecological equivalents of the hedgerows one sees in Europe.

Education

The problem here is that ecological concepts are not always easy to understand, even by those with some scientific education so it is pointless to talk to an illiterate man about primary production or nitrogen cycles but erosion can be seen by anyone and the disappearance of trees is not difficult to appreciate either. I am sure that village elders, even in the most unsophisticated places, are aware and worried about what is happening to their environment, they need every encouragement to do something about it.

Meanwhile the teaching of elementary ecological principles must continue to form part of elementary university education and be included in school curricula. But it is not enough to talk, schools (and universities) must practise conservation as well as teach it.

Literature cited

Beadle, L.C. (1974) The Nuffield Unit of tropical animal ecology. *Journal of Zoology*, *173*: 539.

Charles, J.R. & Keay, R.W.J. (1960) Assessment of the Olokomeji fire control experiment after 28 years. Nigerian Forestry Information Bulletin, No. 3 and summarised in Hopkins B.J. (1965) *Forest and Savanna*. Heineman, London.

Dwyer, P.D. (1975) Observations on the breeding biology of some New Guinea rodents. *Australian Wildlife Research 2*: 33 - 45.

Fitter, R. (1970) Living space for lemurs. *Oryx*. *10*: 365 - 367.

Griffiths, M. (1965) *Echidnas*, Pergamon Press, Oxfor.

Schultze-Westrom, Thomas (no date). *Final report on the 1970 World Wildlife Fund Mission on Conservation in New Guinea*.

SOME ASPECTS OF ECOLOGY AND CONSERVATION OF MOTHS

By

J.J.H. Szent-Ivany, Adelaide, S.A.

In a short address of 10-15 minutes it is impossible to go into general details of the ecology of non-papilionoid Lepidoptera which represent a large number of ecologically, ethologically and systematically different families. So I decided to choose certain aspects of relation between moths and their host plants as the subject of the present talk. In this I include

a/ intraspecific differences in food plant selection and rejection;

b/ observations on seemingly very palatable but in fact detrimental food plants offered to larvae in the laboratory;

c/ a few examples of phytophagia and sarcophagia in congeneric species together with an example of phytophagous and predatory habits within one single species;

d/ the feeding on plants of very distantly related families by morphologically and taxonomically very closely related congeneric species.

Finally I will say a few words on the necessity for protection of the optimal habitats of rare, localized insects, including some moths.

It was more than 55 years ago when I reared my first moth in a breeding jar. This was the saturniid *Eudia pavonia* (Linnaeus) which later presented one of the most striking examples of intraspecific difference in host plant selection.

Merz (1959) reported that in the case of three different varieties of *Eudia pavonia* (one of which developed a distinct race with constant morphological characters) - the decisive factor in the rejection of some food plants was a prolific growth of shield-hairs on the leaves.

Populations in Central and Southeastern Germany, (in the areas of Wurzburg, Frankfurt am Main and Coburg) belonging to the nominotypical *Eudia pavonia pavonia* (Linnaeus) do not favour the elaeagnaceous plant *Hippophae rhamnoides* the leaves of which are covered with a dense growth of shield hairs. When larval populations in this area of Germany were transferred from the rosaceous Crataegus sp, to Hippophae rhamnoides, they have not even touched the leaves of this plant and died shortly after having been transferred.

The females of the nominotypical subspecies further South in the European Alps were observed laying their eggs on the leaves of Hippophae rhamnoides and there the larvae could feed and reach maturity on the elaeagnaceous plant. Thus, the populations in the Alps represented another biological variety of the nominotypical subspecies. The populations in Southeastern Germany and in the Alps do not show any morphological differences.

Larvae of the populations of another race: Eudia pavonia meridionalis Galb. readily feed on the leaves of Hippophae rhamnoides in the Engadin area but they also feed on the leaves of Elaeagnus which have even more rough surface than Hippophae; it has a denser shield hair growth on its leaves. Merz (op.cit.) stressed that the main obstacle which causes the rejection of Hippophae by the northern biological variety of E.p. pavonia is definitely the presence of the thick shield hair growth on the surface of the laminae, not the hardiness of the leaves which might be an obstacle for the tender first instar larvae of some other Lepidoptera. He proved this by the fact that Eudia pavonia larvae in Germany happily fed and grew on the hard and tough leaves of the ericaceous plants Arctostaphylos uva ursi, Calluna vulgaris (the heather) and Rhododendron hirsutum which lack shield-hair on their leaves. E. pavonia is a eurythermal and eurytopous species which occurs from sea level to 2000 metres in the Alps. I reared it in Central Europe on Crataegus and Rubus.

Another striking case of intraspecific variation in relation to host plants has been reported by Paine (1964) who studied the ecology and ethology of the Pyraustine pyralid Nacoleia octasema (Meyrick) which is known under the popular name of banana scab moth, found that populations of this moth do not feed on banana but on various other monocotyledonous plants in mainland New Guinea, Malaya, Sumatra, Borneo, Sulawesi and the Moluccas, whilst populations of N. octasema feed on and cause damage to banana in the Bismarck Archipelago, in the British Solomon Islands, in the New Hebrides, in New Caledonia,Fiji,Tonga, Samoa, on Wallis and Futuna Islands, in North Queensland, on the Lesser Sunda Islands and on Java. Eugene Munroe (Entomological Research Institute, Canada Department of Agriculture), a world authority of the Pyraustinae, who thoroughly examined the morphology of specimens of both: the banana-feeding and the non-banana-feeding populations, made a very careful study of the genitalia of many specimens but was unable to find any differences. Undoubtedly the banana-feeding and non-banana-feeding populations of N. octasema represent two distinct ecological or biological varieties, one could even call them "biological or ecological races" on account of the fact that they are geographically isolated.

Not every plant which is palatable to a moth larva is a useful or nutritive food plant. - in fact some plants offered to certain moth larvae under laboratory conditions in replacement for their natural host plants in the field appear to be extremely palatable but eventually become detrimental to the larvae and they are unable to complete their metamorphosis. Ihad a few such observations during my 17 months of study of the biology of moths, mainly geometrids at the Wau Ecology Institute. Because of shortage of time I only mention here two such cases. Some others will be discussed in my forthcoming papers on the results of my breeding trials.

The larvae of the common Papuan New Guinea cacao and coffee pest, _Ectropis sabulosa_ Warren, a very polyphagous geometrid - is generally known to be mainly a leaf-eater. During my breeding trials I found close to my laboratory a second instar larva of this looper in the centre of a medium sized pink rose-flower where it was chewing the petals. I transferred the larva into a breeding jar and continued feeding it with the petals of the same pink flowered rose variety, not giving it any leaves whatsoever. The _Ectropis_ larva fed with a voracious appetite, produced a very large number of soft pink excreta, much more in 24 hours than other larvae of _E. sabulosa_ fed in other jars and cages on the leaves of coffee and other dicotyledonous plants. It grew much faster than the others. The time interval between the ecdyses was considerably shorter than those of the leaf-eating specimens. However, the larva never became an adult moth. It pupated in loose soil and died in the pupal stage.

A similar example was presented by another Ennomine geometrid which I found on a small _Euodiella_ bush in mossy forest close to the summit of Mt. Kaindi. This rutaceous genus is known only from Queensland and New Guinea and it has two described species (Willis 1966). As I could not find any other bush of its kind in the vicinity I dug out the _Euodiella_ bush and planted it near the laboratory at the Institute's arboretum.

Many larvae which I brought down from the mossy forest zone of Mt. Kaindi (2200 - 2350 metres) to my laboratory at Wau (1150 metres) became very sluggish after a short time; they stopped feeding and died. The reason of this is undoubtedly the altitudinal and climatic difference. The geometrid found on the *Euodiella*, ate the leaves of the small bush planted in the arboretum but it has not eaten a great quantity of leaf; it grew rather slowly - a phenomenon which I also observed on other high altitude geometrid larvae reared under natural conditions, close to their natural habitats. A few days after capture the larva found on the *Euodiella* bush shed its skin. Unfortunately, the apparently high altitude *Euodiella* species disliked the warmer climate of Wau and it wilted away soon after the first ecdysis of the larva. I made a special trip to the area where I found the larva; it was a heavily shaded area, close to the mound of a bower bird. A very thorough search for some more bushes of the *Euodiella* species was unsuccessful. As *Euodiella* belongs to the family Rutaceae, I offered to the larva leaves of two species of *Euodia* which were growing in the J. & M. Sedlacek arboretum of the Institute. It crawled around slowly on the leaves but it had not touched them. Both the *Euodiella* and the *Euodia* leaves have the fresh odour of citrane. Then I offered leaves of mandarin (*Citrus nobilis*) to the larva. These have an even stronger odour of citran than the leaves of the two other plant genera. The larva began feeding on the mandarin leaves immediately and behaved very similarly to the *Ectropis sabulosa* larva feeding on the rose petals. It ate a great lot of leaf material in a short time producing twice as many excretum-pellets as at the time when it fed on its natural host plant. It reached the last instar stage very fast, it pupated in lose soil and - as the rose petal feeding *Ectropis* - it died in its pupal stage.

As for phytophagia and predatorism in congeneric moth species and within one single species, I mention a couple of examples from the Australian, Papuan and Central South Pacific Island regions.

A species of the heliodinid moth genus *Stathmopoda* was found feeding on ripe fruit of *Elacourtia ramonchi* in New Britain (Froggatt 1940) and another species on coconut spathe in New Guinea and the Solomon Islands (Lever 1969a). A third species of *Stathmopoda* was recorded by Common (1966: 44) preying on scale insects of the genus *Eriococcus* in Australia and a fourth species, also in Australia was reared from rust galls of Wattle, *Acacia* (Common, op. cit.) Here we have four closely related taxa, one of which feeds on a dicotyledonous plant, the other one on a monocotyledon, a third one on a cryptogam and the fourth is predacious on other insects.

A striking case of diversity of feeding habits is demonstrated by the cosmopterygid moth *Batrachedra arenosella* Walker the" lesser coconut spike moth" of Papua New Guinea. Populations of this species were found to be phytophagous in Papua New Guinea (Simmonds 1938; Dun 1951; Dumbleton 1954) and on some other islands of the South Pacific (Hinckley 1965; Lever 1969b) where they cause damage to coconut inflorescence * whilst populations of the same species in Australia were found to be predacious. The larvae of the Australian populations feed on scale insects, mainly coccids which are found on bark (Common, op. cit.: 58). A neotropical species of *Batrachedra* has been reported causing considerable damage to the inflorescence of *Cocos nucifera* and *C. romanzoviana* in Brasil (Lever 1969b: 64).

Another interesting phenomenon in the host plant selection of moths is the fact that some morphologically very closely related oligophagous species of the same genus feed on plant species which represent unrelated or very distantly related plant families. In some cases, however, there appears to be certain biochemical affinities between the systematically unrelated or very distantly related host plant families.

Two closely related oligophagous tiger moths (Arctiidae) which have a wide area of distribution and occur also in Papua New Guinea, feed on two systematically very distantly related plant families. One of them *Utheteisa lotrix* (Cramer) the larva and pupa of which was first described and depicted by Howarth (1949, pl. XIV, fig. 2 - 4), feeds on leguminous plants, such as *Tephrosia candida* (Howarth, op. cit.), *Crotalaria, Vigna* and *Desmodium* spp. (Dammerman 1929; Roepke 1941; Kalshoven 1950-51; Szent-Ivany 1956 (1958); Szent-Ivany, in prep.). The other one, *Utheteisa pulchelloides* feeds on boraginaceous plants such as *Heliotropium indicum*, *H. europaeu*, *Tournefortia argentea*, *Messerschmidia argentea* Fletcher 1914: 359; Sevastopulo 1945; Gressitt 1954 : 147 - 148, fig. 65c; Comstock 1966: 16). A third closely related species, *Utheteisa pulchella* (Linnaeus) which does not occur in the South East Asian and South Pacific Archipelagos but is widely distributed in mainland Asia, mediterranean Europe and Africa, is also oligophagous on Boraginaceae, such as *Anchusa*, *Borago*, *Echium*, *Myosotis* (Hruby 1964: 788; Forster & Wohlfahrt 1966: 28). For those who study insect migration, the names *Utheteisa pulchella* and *U. lotrix* are not unfamiliar; both these arctiids are typical "migrant moths" (Hudson 1928, 1950; Forster & Wohlfahrt 1966; Perrott 1966; Williams 1971.)

---

* *B. arenosella* caused very appreciable loss of male flowers of *Cocos nucifera* and nearby smaller islands, "but there was no indication that these losses were large enough to influence fruit setting. "(Department of Agriculture, Stock and Fisheries 1966). The late Hubert Simmonds reported that he "counted 284 entrance punctures of these larvae on the sheaths out of one spike (*Cocos nucifera*) on Bougainville." (Simmonds, op. cit.)

The only reference to a biochemical affinity between Leguminosae and Boraginaceae which I was able to trace in the Literature, was by Plouvier who isolated the mono-ethyl ether L-Borsanitol from 10 species of Lathyrus (Leguminosae) and from plants of 15 boraginaceous genera four of which have been recorded as food plants of U. pulchella in Europe.

Finally I wish to say only a few words on conservation of moths. The chief title of part 9 of this symposium is "Wildlife". Until very recently most research institutes which are concerned with wild life in any country of the world, were mainly interested in the study of the ecology, ethology and conservation of vertebrates, especially mammals and bird. Invertebrates, including insects were rather neglected, although some organizations - other than institutes of wild life research - were responsible for the individual protection of some insects and their habitats for several decades in the United Kingdom and in some countries of continental Europe, such as Germany and Czechoslovakia. The first step in this field in the tropical zone of the Old World part of the Southern Hemisphere was made in Papua New Guinea when seven Ornithoptera species (Lepidoptera: Papilionidae) were declared "protected fauna". The relevant ordinance was gazetted in October 1968 two years after my recommendation was forwarded to various leading authorities and organizations (see also Szent-Ivany & Carver 1967). More recently the Government of Queensland individual protection to the only temperate climate race of the Common Green Birdwing, Ornithoptera priamus (L.) and the Queensland race of Papilio ulysses (L.)

There are also some rare and beautiful moths in Papua New Guinea which have a very localized distribution. Such are the mainly day flying species of the large geometrid genus Milionia. A large percentage of the more than 90 described species of this genus, is restricted or has only been found in Papua New Guinea. Some species appear to be extremely localized and on account of quite a few blank white spots on the entomological map of Papua New Guinea, I have little doubt that new species will be found in the future when the fauna of the areas which have not been visited to the present date by entomologists, will be studied. Further investigations are needed on the distribution and population density of these and other rare and beautiful moths for the purpose of considering their protection.

However, individual protection on its own is not enough. The most favoured (optimal) habitats of Papua New Guinea's rare and beautiful insects should be declared national parks, reserves or sanctuaries, pending on the size of the areas. When such areas will be selected, advice should be sought from both vertebrate and invertebrate zoologists and botanists.

During my recent visit to Hungary (August to October 1975) I had the opportunity of studying the work of the Hungarian National Office of Nature Conservation. One of the latest actions of this office relates to the individual protection and the protection of the habitats of a tiger moth which was believed to be extinct for many decades and which was recently recovered in dense populations in a relatively small swampy area. (L. Gozmany, pers, commun.; Z. Rakonczay, pers. commun., K. Toth, pers, commun.)

Part of my paper published in the 1966 Annual Report of the Papua New Guinea Public Museum and Art Gallery was concerned with the protection of selected blocks as reserves in newly opened large agricultural areas for conservation of both beneficial insects (parasites and predators of pests which may appear in the agricultural blocks planted under monoculture) and rare and beautiful insects. This will be discussed in some detail in my second address prepared for this symposium, entitled "Economic insect ecology and conservation".

****

R E F E R E N C E S.

Common, I.F.B. 1966. Australian moths. Jacaranda Pocket Guide. Second revised edition. The Jacaranda Press, Brisbane. 129 p., 123 fig.

Comstock, J.A. 1966. Lepideptera of American Samoa with particular reference to biology and ecology. Pacific Insects Monogr. 11: 1-74, pl. I-VIII, 1 text fig.

Dammermann, K.W. 1929. The Agricultural Zoology of the Malay Archipelago. Amsterdam. XI + 423 p.

Department of Agriculture, Stock and Fisheries 1966. Economic Entomology. Rep. Dep. Agric. Stk. Fish. Papua New Guinea, 1964 - 65: 107 - 115.

Dumbleton, L.J. 1954. A list of insect pests recorded in South Pacific Territories. Tech. Pap. S. Pacif. Commn., 79 1 - 196.

Dun, G.S. 1951. Some pests of economic crops in Papua New Guinea. Prepared for South Pacific Commission Project E 9., March 1951. List No 5, Schedule 1 a, b. (Cyclostyled).

Fletcher, T.B. 1914. Some South Indian Insects, and other animals of importance considered especially from an economic point of view. Government Press, Madras. 565 p., 439 fig, 50 col. pl.

Forster, W. and Wohlfahrt, T.A. 1966. Die Schmetterlinge Mitteleuropas. Frankische Verhandlung, Stuttgart.

Froggatt, J.L. 1940. Entomologist's Report. In: Annual Report of the Department of Agriculture for the year ending 30th June, 1940. New Guinea agric. Gaz., 7 (4): 298 - 300.

Gressitt, J.L. Introduction. Insects Micronesia. 1. 257 p., 70 fig.

Hinckley, A.D. 1965. Trophic records of some insects, mites and ticks in Fiji. Bull. Dep. Agric. Fiji 45, 116 p.

Howarth, T.G. 1949. Prison camp entomology in the Far East, 1941 - 45. Proc. Trans. S. Lond. ent. nat. hist. Soc., 1947 - 1948: 94 - 110.

Hudson, G.V. 1928. The butterflies and moths of New Zealand. Fergusson & Osborn Ltd. Wellington. 386 p., 67 pl.

Hudson, G.V. 1950. Fragments of New Zealand Entomology. Fergusson & Osborn Ltd. Wellington. 188 p.

Kalshoven, L.G.E. 1950-1951. De plagen van de cultuurgewassen in Indonesie. I., II. N.V. Uitgeverij V. van Hoeve's - Gravenhage. 1066 p., 16 pl., 579 fig.

Lever, R.J.A.W. 1969a. A check list of economic insects and mites of crops in the British Solomon Islands. FAO Plant Protection Committee for the South East Asia and Pacific Region Technical Document No. 65. 12 p.

Lever, R.J.A.W. 1969b. Pests of the coconut palm. FAO Agric. Studies. 77, XIII + 190 p. 106 fig.

Merz, E. 1959. Pflanzen und Raupen. Uber einige Principien der Futterwahl bei Grossschmetterlingen. Biol. zbl., 1: 152 - 188, 2 fig.

Paine, R.W. 1964. The banana scab moth Nacoleia octasema (Meyrick), its distribution, ecology and control. Tech. Pan. S. Pacif. Commn. 145, 70 p., 21 pl.

Perrott, D.C.F. 1966. Records of the Arctiid genus Utheteisa in New Zealand. N.Z. Ent., 3 (5): 3 - 4.

Roepke, W. 1941. On the Javanese species of the genus Utheteisa Hb. (Lep. Het., fam. Arctiidae). Tijdschr. Ent. 84: 1 - 9, 4 fig.

Sevastopulo, D.G. 1945. The early stages of Indian Lepidoptera. Part XI. J. Bombay nat. Hist. Soc., 43 (3); 309 - 415.

Simmonds, H.W. 1938. Coconut pests and diseases in Melanesia and Northern Polynesia. Bull. Dep. Agric. Fiji 20: 1 - 40, 4 pl.

Szent-Ivany, J.J.H. 1956 (1958). Insects of cultivated plants in the Central Highlands of New Guinea. Proc. Inern. Congr. Ent., 10 (3): 427 - 437.

Szent-Ivany, J.J.H. 1967. Insect collection and insect protection. In: Annual Report of the Trustees of the Public Museum and Art Gallery of Papua and New Guinea for the year 1966: 18 - 24, 2 fig.

Szent-Ivany, J.J.H. Economic insects of Papua New Guinea. (In preparation.)

Szent-Ivany, J.J.H. & Carver, R.A. 1967. Notes on the biology of some Lepidoptera of the Territory of Papua and New Guinea with the description of the early stages of Ornithoptera meridionalis Rothschild. Trans. Papua New Guinea scient. Soc. 8: 3 - 35, 2 fig., 10 pl.

Williams, C.B. 1971. Insect Migration. The New Naturalist Series. First edition 1958, reprinted 1971. Collins, London. 237 p., 11 col. illust., 22 phot. illust.

Willis, J.C. 1966. A dictionary of the flowering plants and ferns. Seventh Edition. Reviewed by H.L. Alry Shaw. Cambridge, at the University Press. 1214 + LIII p.

****

THE MOTH FAUNA OF PAPUA NEW GUINEA.

A. P. Mackey.
Biology Dept. UPNG.

INTRODUCTION.

Investigations of the moths of P.N.G. began when A.S. Meek started collecting specimens for the Hon. Walter Rothschild in the early 1890's. He discovered that many rare and beautiful moths could be caught by attracting them to large camp fires built in rain forest clearings. During the 1890's he made moth collections from several areas in P.N.G, mainly around Milne Bay, but also around the head-waters of the Aroa River in the Owen Stanley Range. Much of what we know of the moths of P.N.G. comes from Meek's collecting, together with the collecting of the Pratts (1902-4), Cheeseman (1930's) and Brandt (1950's). Although others have collected in various areas, until recently, these four people were the only ones to have collected systematically and at length. Figure 1 shows the approximate areas where collecting has been carried out in P.N.G. As can be seen collecting has not been carried out in large areas of the country and much work remains to be done.

THE FAUNA.

Very little is known of the moths of P.N.G. The early collections have been worked through taxonomically, but as yet there is not even a complete list of those species known to occur here. The taxonomic work completed so far, with the exception of a little modern work, is not of good quality - species descriptions are poor and in most cases cannot be used to identify specimens. There is also the added difficulty that this information is in journals difficult to obtain in this country. The available distribution data on those species known to occur here are scattered throughout the literature and most remains unpublished and has to be obtained from data labels on museum specimens. Data thus obtained is often unreliable, as the early naturalists were not renowned for the accuracy with which they recorded their localities or habitats. With such an uncertain background, recent moth studies must inevitably start from scratch. The first priority must be to produce a species list for each major habitat, together with workable species descriptions as a basis for future work.

The moth fauna of P.N.G. is extremely diverse (a conservative estimate for the larger moths would be about 2000 species) and contains many beautiful and spectacular species. For example, the largest pyralid in the world is found here and the largest moth in the world, Coscinocera hercules is common is some areas of lowland rain forest. Many of the smaller Geometridae and Notodontidae are brilliantly coloured and intricately patterned. From what is known, the fauna is mainly Oriental in origin showing many affinities with the fauna of the Malay Archipelago.

So far, the four major habitat divisions in P.N.G. have been sampled i.e. lowland rain forest, montane forest, alpine grassland and savanna. Each contains a distinctive moth fauna, but so far, detailed comparisons of these faunas have not been made, although it is known that the fauna of the savanna is very similar to the Australian savanna fauna, and that there are substantial differences between moth faunas from very similar habitats. The moth fauna from lowland rain forest around Kiunga is quite different from that of lowland rain forest around Angoram.

ECOLOGY.

The richness of the moth fauna leads on to considerations of moth life histories and ecology. Very few life histories are known and even less is known of general ecology - how species are distributed laterally and vertically, whether populations show seasonality and whether they are stable or show large fluctuations. The functional role of moths in tropical forest ecology is also not known - e.g. are moth larvae important in energy flow, and are the adults capable of fertilizing the flowers of the many widely dispersed plant species which occur in tropical forests; can moths fly far enough?

Modern ecology suffers from the lack of systematically collected, sizeable samples of animal populations. This is particularly true of tropical environments which are rich in species and thus need large samples to be taken for general studies. Moths are easily sampled in large numbers by light trapping and are relatively easy to identify. They are thus ideal for general studies in tropical ecology - such things as how is high species richness brought about, and how is it maintained? Aspect diversity in P.N.G. moths is high, suggesting higher predation in tropical forest than in temperate forest, leading to high moth diversity. Species richness indicates much competition within the Lepidoptera; is this due to reduced niche volume brought about by habitat partitioning, temporally (different flight times and periods) or spatially (flight stratification). This would suggest diffuse competition, which would be most severe between phenotypically similar species. Thus small families tend to be represented by species with high population densities. In this respect it is interesting to note that on Mt. Kaindi some of the most abundant moths belong to the large family Noctuidae, of the genus *Stictoptera*, the species of which are renowned for their intra-specific variation. *S. cuculloides*, the commonest on Mt. Kaindi, has 7 morphs recognized from there so far.

CONSERVATION.

To date, so little is known of the moths of P.N.G. that we do not even known what there is to conserve: Are there any endangered species that need to be conserved? What is needed however, is to conserve areas of each habitat to ensure that future research on the sorts of problems outlined above can be carried out. P.N.G. is rich in having many different, undisturbed types of habitat: It has the largest undisturbed area of tropical rain forest in the world.

By conserving these habitats, P.N.G. will be helping herself and the world scientific community, by providing areas in which important research in tropical ecology can be carried out. She will also be helping those countries that have already left it too late to conserve their own habitats.

ASPECTS OF BUTTERFLY CONSERVATION IN PAPUA NEW GUINEA

T.L. Fenner,
D.A.S.F., Konedobu.

Firstly, I would like to nominate butterflies as a key group for insect conservation, not so much in ecological terms but for psychological reasons.

I think we must all agree that conservation needs to be sold to people in general and decision-makers in particular. This is by no means an easy task, and we need all the help we can get! Surely butterflies, which have a good image because of their beauty, pleasing behaviour and relative non-pest status, are an ideal group of insects to attract public interest in and sympathy towards insect conservation in general. Perhaps even more importantly, a good number of species have measurable cash value as collectors' specimens, and this is the sort of talk which quickens the pulses of decision-makers and their economic advisers. I offer the following estimate for just one species : the common green birdwing *Ornithoptera priamus poseidon* (not a protected species). The world market could probably absorb 5000 pairs per annum. At K4 per pair, this represents a gross return of K20000 in foreign currency. At the individual level, a competent collector in a suitable area should have little difficulty in selling 125 first quality pairs per annum. This might mean an average harvest of one pupa per day, which allows for rejection of almost one third of collected material and is a harvest level which should have no deleterious effect on the wild population. In economic terms, the collector's gross return would therefore be K500 per annum. Deduction of K35 for expenses (mainly air mail postage) leaves a net income of K465, which slightly exceeds the present rural basic wage.

Commercial collection and sale of this one species could therefore theoretically sustain forty rural Papua New Guineans who, with their familes, might total two hundred people - surely a most worthwhile contribution to national development!

To return to butterfly conservation in broader general terms, it is, of course, true for all organisms that the amount and quality of available habitat determines where, when and in what numbers they occur, and protection of habitat is therefore the key factor in conservation of insects, including butterflies. For protection of regional faunas, conservation of large tracts of a wide range of habitat types ranging from mangrove flats to high montane shrubland is needed. These extremes and all intervening habitats support distinctive butterfly faunas.

Highest priority should be given to offshore islands according to their size, human population and isolation. The Trobriand Islands, for example, have at least 70 species, including two endemic species and many endemic subspecies. On the main island, which supports a dense human population, very little relatively undisturbed forest remains. On a week's visit there in good weather in 1973, I was unable to locate 23 (i.e. one third) of them, despite prolonged searching in many habitats.

Other relatively small islands with significant endemic butterfly faunas include the D'Entrecasteaux, Louisiades, Woodlark, Kar Kar, Admiralties, New Hanover, Witu and probably St. Matthias (Mussau) and the Duke of Yorks. Habitat conservation on all of these should have very high priority. High montane areas, which may also be "ecological islands", are at somewhat less risk because they are generally less disturbed by human activity.

Over their entire distributional range, the larvae of most butterfly species are oligophagous, utilizing a few (usually related) plants as food sources. Within any given locality, only one larval foodplant may occur, in which case the butterfly is locally monophagous. Adult food sources include nectar, fermenting sap and/or fruit, urine, faeces and water containing dissolved salts and organic compounds. Not all of these sources are likely to be used by any one species in any one area, and there often seem to be distinct preferences for one type of food source at the species level and sometimes at the species/sex level. Males of certain *Graphium* species, for example, commonly drink from puddles whereas females are never found at such places and presumably feed at flowers.

Other habitat requirements are less well known, but do not seem to be as critical as food sources. Also, many forest butterflies are sedentary, tending to remain within an area of a few hundred square metres throughout their adult life. Such places are the "spots" referred to by lepidopterists in which a particular species can regularly be found.

Because of the foregoing, and also because the life histories of many species are not known in Papua New Guinea, careful baseline surveys of proposed conservation areas (both on islands and the mainland) are essential to ensure that they include adequate areas of habitat for populations unique to or characteristic of the region.

It should also not be overlooked that even in quite highly developed areas pockets of suitable habitat as small as a hectare or two may be capable of sustaining populations of some butterfly species on a long-term basis, especially if a number of such refuges occur within an area of a few square kilometres. In some parts of the world comparably small refuges represent the sole remaining habitats of some butterfly populations.

I would like to return now to another attribute of butterflies which was touched upon in my opening and has very important implications in their conservation.

More than any other group of organisms except, perhaps, sea shells, they have the misfortune to be "collectors' items" in the classical sense of the term.

This has at least 2 important consequences especially by comparison with the vast majority of other insects.

1) Although, as previously stated, conservation of habitat is the paramount factor influencing their chances of survival, there is a distinct threat that if the total area of suitable habitat of a population or perhaps an entire species is very limited, sheer collecting pressure could reduce the population to a critical level for its survival - even if the habitat remains otherwise suitable.

   This risk is increased by the unfortunate fact that the harder to get (e.g. rarer or more strictly protected in this example) a "collector's item" becomes, the more desired it is - therefore the higher its price and therefore the greater the incentive to commercial interests to obtain it.

   This is likely to encourage over-exploitation ("Get 'em while we still can!"), which completes the vicious cycle by depleting wild stocks still further. A sad state of affairs indeed, but a common enough one in the annals of conservation!

2) There may therefore be sound conservational and economic reasons for culturing or farming certain species to adequately supply the market with good quality specimens, thereby easing pressure on remaining wild populations.

Perusal of dealers' catalogues and trade journals such as "Insektenborse" indicates that there is a solid and continuing demand for certain groups of butterflies, among which birdwings figure prominently.

Another inescapable point is that even the largest butterflies are relatively small, and therefore easily concealed and smuggled. For example, a good-sized and carefully papered pair of the world's largest species, Ornithoptera alexandrae, could be adequately fitted into an Australian Passport wallet - although the bearer would need to carry his passport somewhere else!

I think that we must therefore accept that a world demand for specimens of birdwing butterflies will continue to exist regardless of whether they are protected or not. Further, I believe that their high cash value and the relative ease with which they can be smuggled will ensure continued attempts to supply the market illegally unless provision is made for the legal supply of specimens.

I would therefore urge most strongly that this objective be thoroughly investigated as a matter of urgency because I believe it would strike at the source of the present trafficking problem by virtually eliminating the illegal market.

This is not the place and I am not the person to discuss the administrative and regulatory aspects of the proposal, but I would like to consider the entomological aspects of how it might be achieved.

For a start, it has become increasingly apparent in recent years that, with the single exception of O. goliath, none of the species is rare in areas of prime habitat at the most favourable times of year. For example, observations of concentrations of adults numbering approximately 30 (O. meridionalis) and 50 (O. chimaera) have been made on single large flowering trees. Even if it is conceded that these could represent the bulk of the adult population from an area of, perhaps, 15 sq. km. of forest, the overlap of generations and estimated fecundity of females would suggest that the area is probably carrying something like ten times as many early stages at various stages of development.

One possibility would therefore be to legalize limited collecting from wild populations for commercial purposes. However, an immediate difficulty is that we know far too little about the ecology of any of the species concerned to even estimate what might constitute an acceptable level of harvest from any given population under a particular set of circumstances. It is ironic that perhaps our most useful guide is limited, but increasing, information on the levels of illegal collection of some species which have occurred in certain areas compared with their present status in the same localities.

There are, however, reasons for considerable optimism about the technical possibilities of a second alternative, namely culturing or farming the range of birdwing species. Most of the protected species appear to be monophagous in any given locality (perhaps throughout their range). Known larval foodplants are all in the genus *Aristolochia*, and the majority have been tested and can be grown from seed or, often, from cuttings.

Furthermore, certain common garden plants are favoured food sources for the adults of a number of species. Such plants include the common red *Hibiscus* and *Ixora* and tulip tree *Spathodea campanulata.*

At the simplest level, habitat enrichment by the planting of favoured adult food sources and the correct larval foodplants in areas of natural occurrence of a species would allow the collection of pupae from the cultivated area. Release of a proportion of the resulting adults may or may not be necessary, depending on the status of the wild population which would, in turn, be largely determined by the amount and quality of natural habitat sustaining it.

Bearing in mind that release of some bred adults is likely to be desirable where the local wild population is relatively small, released individuals should NOT consist solely of small-sized or malformed specimens, as these phenotypes would presumably reflect a combination of poor adaptation to the culture environment and possible genetic disadvantages. On the other hand, release of a balanced proportion of the large and vigorous adults obtained should benefit the culture operation. Such individuals, presumably genetically fit and well adapted to culture conditions, should swing the attributes of the wild population toward fitness for the culture environment.

A more sophisticated technique, and one which places for fewer demands on a sustaining wild population, is the culturing of successive generations of a species in captivity - silkworms being a classical example amongst insects. Although birdwings are not quite so amenable to domestication, they have some important desirable attributes for it.

(1) Feeding presents few problems. In addition to what was said earlier, it is worth noting that adults are easily hand-fed (e.g. on honey-water mixture) in captivity. Artificial diets for late stage larvae of those species which feed on woody tissues could probably be developed and, further, it is known that larvae of at least two protected species will accept the most common, hardy and quick-growing native Aristolochia species (A. tagala) and complete their life cycle on it, even though females are not known to oviposit on that host in the field.

(2) Mating can take place in the field very soon after emergence of the female, before her wings have fully hardened.

Hand-mating can also be carried out successfully with some expertise.

(3) Females of at least some species will oviposit in large sleeves or reasonably small cages. However, this aspect of farming is likely to be the most limiting factor and needs urgent study.

(4) Although eggs are sometimes parasitised by microhymenoptera and eggs and young larvae may be preyed on by various arthropods, natural enemies are unlikely to be a serious problem in captivity. Larvae also show little susceptibility to disease, but fighting may be a problem under crowded conditions.

(5) Some practice is needed in handling pupae and newly emerged adults in order to obtain a high percentage of first quality specimens. Maintenance of continuously high, but only rarely saturated, humidity is important, as is avoiding disturbance of teneral adults before the wings are adequately hardened - usually some six hours after emergence.

It must be stressed that there are still many unknowns and possible pitfalls, and much more detailed investigation is needed before any ambitious attempts at mass culturing are contemplated. An institutionally operated pilot rearing project supported by full laboratory studies would probably be the best approach, and would need high levels of both scientific and practical expertise to achieve its aims. Virulent infectious diseases, for example, are a major hazard of insect mass cultures, and pathological aspects should be very closely checked.

All in all, I believe that there are promising prospects for farming some and possibly all of the protected birdwing species. This would have the conservation benefit of knocking the bottom out of the illegal market by guaranteeing availability of first class specimens at reasonable prices, and at the same time earning foreign exchange for the country and income for the farmers - surely both desirable in terms of national development goals.

BIOLOGICAL CONTROL

by

J.P. Spradbery

CSIRO Screw-worm Fly Project,
P.O. Box 6712
Boroko
Papua New Guinea

Since man appeared on earth, he has broken nearly every ecological principle of population control, community interraction and energy flow. He has mismanaged land, forests and water and spread diseases and alien forms of life while extirpating endemic animals and plants. More recently man has reproduced at such a rate that he has forced too many of his own kind into ecosystems that cannot absorb them. With the development of man's system of monoculture crop production, he has so disorganised the ecological balance that has taken millions of years to achieve that he has created unprecedented pest problems for himself and his agriculture.

Three decades ago, with the development of DDT and similar broad-spectrum pesticides, it was thought by many that man was on the brink of winning the war against those organisms which compete for his health and supplies of food and natural materials. However, the realisation that pesticides frequently aggravate the situation by their non-selective action, create residue problems which disrupt important food chains and frequently lead to resistance in the target pest has led to a resurgence of interest in the use of biological agents and techniques in pest management systems.

Classical biological control is defined as the manipulation of natural enemies to limit the abundance of plants or animals. Implicit in this definition is the conservation of the habitat and its rich variety of plant and animal life and an acknowledgment that the endemic fauna is a useful natural resource for biological control. Although the more striking and well known examples include the control of pests and weeds by the use of predatory, parasitic and phytophagous insects, other organisms such as microbial diseases, nematode worms and fish have been employed inthe control of pests. A contemporary example is the control of *Sirex* wood-wasp in the exotic pine forests of Tasmania and Victoria. *Sirex* kills trees by injecting a symbiotic fungus and phytotoxic mucus into the wood at the time of oviposition but causes little damage in the Mediterranean countries where it is endemic. As with many introduced pests, *Sirex* became established without the complement of natural enemies which keep it below economically acceptable levels in its area of origin. CSIRO established a laboratory in Europe to investigate *Sirex* ecology and search for natural enemies in places where the climate was analagous to those areas in Australia where *Sirex* presented a problem. A suite of parasitic insects and nematodes was discovered which were consigned to Australia for culture, study and subsequent release. Because the introduced enemies of Sirex rely on the woodwasps' highly specific fungus either to locate their hosts or provide food for the free-living stages of the nematodes, there was no danger of these introductions causing deleterious side-effects. The intensive ecological studies carried out in Europe and Australia and the success in establishing natural enemies in Australia as well as sound forest management practice has led to a significant reduction in Sirex and to averting a possible catastrophe in the soft-wood forests.

The biologist's armoury in his war on pests has in recent years been augmented by a variety of innovations coupled with an increasing sophistication in ecological research, especially population studies. For example, the detection, assessment of biological activity and synthesis of insect pheromones (essentially sex-attractants) has led to new methods of control and means of determining the abundance of many pests by selectively trapping the responsive sex. The current CSIRO work on the biological control of cattle dung in Australia illustrates a novel approach to an agricultural problem. Cattle were introduced into Australia without those organisms which utilize large dung pads and, with the arid climate of much of the continent, cattle dung may remain on the ground for five years, rendering up to four per cent of the grazing area unproductive each year. Africa is the home of large herds of animals together with a range of organisms adapted to feed on their dung and it was here that CSIRO began to search for dung destroyers. The programme involves the introduction of beetles which break down the dung, thereby aiding soil fertility and disrupting the development of noxious insects, such as bushfly and buffalo fly, which inhabit the dung. Additional beetles which are directly predatory on the larvae of these flies are also being introduced. Some of these African beetles are now being evaluated in Papua New Guinea to determine whether they can establish themselves and prove effective in their role as dung dispersers.

The development of techniques for sterilizing insects has given rise to one of the most dramatic successes in biocontrol, that of screw-worm fly eradication in parts of the U.S.A. Screw-worm fly maggots feed on the flesh of living, warm-blooded animals, causing great destruction in livestock. To control this pest, flies were produced in vast numbers using artificial rearing methods, sterilized by gamma irradiation and released as adults from aircraft over infested area. The theory of the exercise was to flood the region with sterile males so that the wild females would more frequently mate with the released males and thus lay infertile eggs. The theory was confirmed in practice and in two years the screw-worm fly was eradicated from the south-eastern United States. Although success with this technique has not been matched with other pests, genetic studies are beginning to stimulate novel approaches to pest management.

Geneticists now have the capacity to develop synthetic strains of organisms whereby lethal traits or inherited sterility may be introduced into pest populations. Similarly, disease-transmitting vector insects could be displaced by non-vector strains or species while the recent synthesis of an oil-feeding bacterium now offers hopes for the biological control of oil spillage at sea.

In pest management, the cost effectiveness of the alternative control measures generally determine the control or containment strategy. Here it must be admitted that biological control investigations demand a considerable expenditure of time, money and skilled effort. However, the annual costs of $70 million to develop new pesticides and sales of $120 million compared to a meagre $4 million invested in biological control displays a quite irrational disbursement of expenditure. Nevertheless, over recent years, there has been a proliferation of international agencies involved in pest control projects, many directed at providing advice and technical assistance to developing countries. With more than twenty such agencies, panels and sub-committees it is hoped that Parkinson's law does not prevail to the exclusion of practical endeavours in biological control ventures.

While promoting biological methods of control, it must be acknowledged that chemical pesticides perform a useful role and, realistically, they should be integrated into a system of pest management which utilizes all the suitable techniques and methods as compatibly as possible to maintain pest populations below the levels at which they cause economic injury. For the implementation of integrated control measures, a thorough understanding of the ecological requirements of the host, the pest and the biotic agents is absolutely crucial.

Global population problems allied to a need to exploit ever more of the earth's natural resources for food and materials will inevitably lead to even more alarming confrontations with pests. With the limitations inherent in chemical pesticides, it is of paramount importance that we support the more permanent, safe and, in the long term, cheaper, expedient of biological control.

BIOLOGICAL CONTROL OF WEEDS

by

A.A. Kirk
Biology Department
U.P.N.G.

INTRODUCTION

The biological control of weeds consists essentially of the limitation of their abundance by the action of their natural enemies. Of the 300,000 species of Angiosperms about 1% are serious weeds, 3% are recorded as weeds and the total number of weeds may be 10-12% (Wilson, 1970).

The number of successful limitations of weeds is small and is proportional to the amount of research carried out and the number and variety of natural enemies used (De Bach, 1964). Some results have been spectacular such as the control of prickly pear cactus by insects and the current control of skeleton weed by a rust fungus *Puccinia chondrillina* Bul & Syd in Australia (Hasan, 1972).

Some advantages of biological control are that the method produces no environmental pollution, no harmful effects on man and that biological control agents are self-perpetuating. Chemical control of weeds will continue to be necessary but it is clear that chemical control should supplement rather than replace natural control.

Work done on Pacific Islands relevant to Papua New Guinea and the strategies that might be employed in the biological control of weeds both introduced and native in Papua New Guinea are discussed.

RESULTS OBTAINED

Table I lists the attempts carried out on some South Pacific Islands, and their results. (Rao et al. 1971, Wilson, 1960).

The attempts made in Papua New Guinea are shown in Table 2 (D.A.S.F. reports 1965-66; 1967-69).

STRATEGIES FOR BIOLOGICAL CONTROL OF WEEDS

A sound knowledge of the ecology of the weed in its native and adopted habitats is essential when assessment of natural enemies is being made. The success of an introduced natural enemy depends on climatic tolerances, host specificity, host selection, effective rate of damage and ability to disperse. Natural enemies already selected for weed control show differing modes and effects of attack.

The natural enemies selected for the control of skeleton weed *Chondrilla juncea* L. in Australia are a good example of this (Wapshere, 1969). An eriophyid gall mite causes galling of the skeleton weed flower and can reduce flowering and seeding by 100%. A gall midge attacks the young rosettes and stems causing extensive dieback. A root moth attacks the deep root of skeleton weed and reduces regrowth in Spring. The other organism currently used in the attack on skeleton weed is *Pucciana chondrillina*, a rust fungus which attacks the leaves and stems leading to extensive damage to the plant (C.S.I.R.O., 1973).

These organisms were all selected for importation in Australia, from the Mediterranean region, after they were shown to be limited in their host range to *Chondrilla* or closely related plants. The selection of plants for host testing following the criteria of Wapshere (1970, 1970a) based on phylogeny.

Testing should be carried out as near as possible in field conditions to ensure normal behaviour of the potential control organism.

Other factors are taken into account when introducing organisms, e.g. an aphid was not used against *Chondrilla* in Australia because it was considered it might have acted as a vector of plant viruses. (C.S.I.R.O. 1973).

In the case of a native or cosmopolitan weed natural enemies may be sought in other parts of the weed's range. *Tribulus cistoides* L. (Table 2) is pan-tropical in distribution and has been controlled in Papua New Guinea by a weevil imported from another part of its range.

There has already been selected movement by man of natural enemies of some insect pests from district to district and from one island to another in Papua New Guinea (Wilson, 1960). With an increased knowledge of the ecology of native weeds, further movements of beneficial organisms may be possible.

A weed currently being studied at the Biology Department of U.P.N.G. is bracken fern *Pteridium aquilinum* (L) Khun. This persistent poisonous weed is indigenous in many parts of the world. It is common in Papua New Guinea where it occurs from 100-3,300 metres above sea level.

In many countries it is a major pest of pastures where it displaces economically valuable grasses. In addition, bracken has been shown to be carcinogenic (Hirono et al, 1973) inducing tumours in cattle and sheep which have eaten it.

Bracken has the potential of becoming a serious weed in Papua New Guinea, as more land is cleared and burned to provide range land for the increasingly important cattle and sheep industries.

Early surveys show that there is a complex insect fauna attacking bracken in Papua New Guinea. The species content of the fauna differs with altitude and it may be possible to help control bracken in one part of the country with insects moved from another.

According to Henty and Pritchard (1973), there are 140 weed species in Papua New Guinea of which 50 are known introductions to Papua New Guinea. Table 3 lists some of the important ones which originate from Tropical America. These wold be candidates for biological control using natural enemies present in their native habitat.

REFERENCES

C.S.I.R.O., 1973. Annual Report of Division of Entomology. 144 pp.

D.A.S.F. 1965-66. Annual Report, p. 120

1967-69. Annual Report, p. 179

De Bach, P. 1964. Biological control of insect pests and weeds. London: Chapman and Hill Ltd. 844 pp.

Hasan, S. 1972. Behaviour of the powdery mildews of *Chondrilla juncea* L. in the Mediterranean. Actas III Congr. Un. Fitopat. Medit. Oeiras: 171-176

Hirono et al 1973. Comparative study of carcinogenic activity in each part of bracken. Journ. Nat. Cancer. Inst. 50(5):1367-1371.

Rao, V.F., et al. 1971. A review of the biological control of insects and other pests in south east Asia and the Pacific region. C.A.B. Farnham Royal, England, 149 pp.

Wapshere, A.J., 1970. The assessment of the biological control potential of organisms for controlling weeds. Proceedings 1st International Symposium on Biological Control of Weeds. C.I.B.C. Misc. Publ.1: 79-80.

Wapshere, A.J., 1970a. The assessment of biological control potential of the organisms attacking *Chondrilla juncea*. Proceedings, 1st International Symposium on Biological Control of Weeds. C.I.B.C. Misc. Publ. 1:81-89.

Wilson, F. 1960. A review of the biological control of insects and weeds in Australia and Australian New Guinea. Tech. Comm. 1. C.I.B.C. Ottawa, Canada, pp.102.

Wilson, F. 1970. The nature and advantages of biological control. Advance. Sci. 26:374-78.

TABLE 1

BIOLOGICAL CONTROL OF WEEDS ON SOME PACIFIC ISLANDS (Rao *et al* 1971, Wilson 1960)

Weed	Origin	Pacific Island	Introduced Natural Enemy	Origin	Control Obtained
Lantana camara L.	Central America	Cook Islands	*Uroplata girardi* Pic. (Hispid)	South	Considerable damage being done
		Fiji	*Ophiomya lantanae* (Frogg) (Fly)	Mexico via Hawaii	Good
			Thecla lazochii (Godart) *T. echion* (L) (Butterflies)	Central America via Hawaii	*T.echion* failed
			Teleonemia scrupulosa Stal (Bug)	Mexico via Hawaii	Only effective in dry weather
			Blepharomastix acutangulalis (GN) *Diastema tigris* (GN) Moths	Mexico	Failed
			Syngamia haemorrhoidalis (Guen) (Moth)	Central America	Established but not effective
		New Caledonia	*O. lantanae* *T. scrupulosa*	Mexico via Hawaii	Well established. Limiting spread
		Trust Territory	*T.scrupulosa* *Epinotia lantanae* (Busck)	Mexico	Limiting spread
			O. lantanae *Platyptilia pusillidactyla* (Wik) (Moth)	Hawaii	Established

TABLE 1 (Cont.)

Weed	Origin	Pacific Island	Introduced Natural Enemy	Origin	Control Obtained
Lantana camara L. (Cont.)	Central America	Norfolk Island	T. scrupulosa	Mexico via Aust.	Limiting spread well
Clidemia hirta (L.)D. Don Koster's Curse	West Indies	Fiji	Liothrips urichi Karney (Thrips)	Trini-dad	Controls by inhibiting growth and reducing invasion
		Solomon Islands	Liothrips urichi		Cut weeding costs by 75% - weed not recovered.
Elephantopus mollis(Kuntl) Tobacco weed	West Indies	Fiji	Tetraeuaresta obscurivent-ris (LW) (Seed-fly)	Trinidad	No estimate made
Opuntia spp. Prickly Pear	South America	New Caledonia	Cactoblastis cactorum Berg (Moth)	Argen-tina	Good
			Dactylopius nr sp.confusus (Ckll) (Coccid)	Argen-tina	Very good
Xanthium strumarium L. Noogoora burr	North America	Fiji	Euresta aequalis (LW) (Seed-fly)	U.S.A.	Not recovered

TABLE 2

BIOLOGICAL CONTROL OF WEEDS IN PAPUA NEW GUINEA

(DASF 1967/69)

Weed	Origin	Introduced Beneficial Insect	Origin	Control Obtained
Tribulus cistoides L. Puncture vine	Pan-Tropical	*Microlarinus cyprifornis* (Walk.) (Weevil)	Hawaii	Good
Lantana camara L. Lantana	South America	*Hypena strigata* (F) (Moth)	East Africa	None

TABLE 3

SOME IMPORTANT CANDIDATE WEEDS FOR BIOLOGICAL CONTROL IN PAPUA NEW GUINEA (from Henty & Pritchard 1973)

Weed	Family	Origin	Disagreeable Effects	Distribution in Papua New Guinea
Paspalum conjugatum Berg Thurston grass	Graminae	Tropical America	Serious competitor to tree crops; coffee/tea	From sea level - 2000 m.
Digitaria insularis (L) Mez. Feather top grass	Graminae	Tropical America	Unpalatable to livestock, formidable competitor.	Markham Valley
Althernanthera pungens Kunth Khaki Weed	Amarantha-ceae	Tropical America	Said to cause deaths of pigs eating it, and Dermatitis in cattle.	Central and Morobe Districts
Hyptis suaveolens (L) Poit	Labiatae	Tropical America	Considered dangerous to livestock	Widely spread at low altitudes
Mimosa pudica L. Sensitive plant	Leguminosae	Tropical America	Suspected of poisoning cattle.	Low altitudes sometimes common

ECONOMIC INSECT ECOLOGY AND CONSERVATION

by

J.J. H. Szent-Ivany, Adelaide, S.A.

It is needless to stress that ecology together with ethology are two of the most important disciplines, we may say essential tools, of the work of the economic entomologist.

Facing a new pest which appeared on an indigenous or an introduced crop we have to study the pest in relation to its environment, if it is a native species of the country: in both the virgin rain forest in its "wild state" and in the cultivated area which it invaded when attacking our agricultural plant or forest tree. We have to study the bionomics, the natural enemies and many aspects of behaviour of the species before we are ready to work out the most feasible, most economical and successful method of control, whether it be chemical, biological or integrated.

As in my address on the ecology and conservation of moths, my time is restricted/This does not allow me to deliver a comprehensive general address on economic insect ecology many aspects of which were very thoroughly discussed by Professor Lamb with special reference to Papua New Guinea's insect pest problems. So-as suggested by Dr. Gressitt - I will say a few words on the correlation or may I use the term synchronization of the interest of both the agriculturist - economic entomologist and the insect-conservationist. This, at first glance sounds rather controversial but I could give more than one example where the interests of the agriculturist and the conservationist meet, not so much in the individual protection of insects as in the protection of certain areas where the protected insects have their densest populations, where they have their optimal habitats.

Time permits me to discuss one such example to which I have already referred in my address on moth ecology and of which I wrote in some detail in a paper written for the Annual Report of the Papua New Guinea Museum to the request of the President of the Board of Trustees, the late Sir Alan H. Mann with whom I worked for some years as one of the Museum's Trustees. (Szent-Ivany 1967a). As probably most of you have not read this paper, I will read the relevant paragraphs and also mention briefly my ideas on the similarity of the concept of the "fair hunter" and the fair insect collector" both of which are fauna conservationists in the true sense of the word. This also is part of my paper published in the museum report and I discuss this subject first.

A few years ago there was an International Hunting Exhibition in a European Capital City. I happened to know one of the main organizers of the Exhibition. He wrote to me asking to draw the attention of Australian scientists: zoologists, ecologists, wild life researchers, conservationists and others to contribute to this exhibition. The first one whom I contacted was a zoologist friend, a fine specialist in his field and a prominent young scientist. "Hunting is a dirty word here - he answered - most vertebrate zoologists are conservationists, not hunters." I had to tell him something that he probably did not know: the fact that the first conservationists who ever did something in the field of fauna preservation were the type of people which we used to call in Europe "fair" or "gentleman" hunters whose aim was not shooting but preserving, maintaining and defending their game. English and German hunters were especially prominent in this field and for present day conservationists who are interested in the history of their subject, it would be most rewarding to read some hunting text-books published in the last century 70-80, 100 or more years ago. In fact, even in our days - many prominent conservationists are at the same time what we may call - "fair hunters".

The fair hunter's golden rule in German reads: "Kein Häger, kein Jäger." This could be translated into English as follows: "He who is not protector (of his game), is not (fair) hunter". As far as I know Australia has not sent any exhibits to the above-mentioned International Hunting Exhibition but I have to mention that one of the largest exhibits shown there was on wild life conservation.

I compare the fair insect-collector with the fair or gentleman hunter. He should try to preserve the rare insects of his country; furthermore he should not collect indiscriminately; he should let loose damaged specimens of a series of harmless non-economic insects of no pest status from which he has a large enough series to study its systematics, individual variation and distribution.

I am convinced that in some cases individual protection of rare and beautiful insects is justified, but - as I said earlier - it is more important to protect certain selected areas where the rare species are the most abundant or relatively abundant.

As for the common interest of the agriculturist - economic entomologist and the insect-conservationist, this occurs in virgin forest areas inhabited by some rare insects and which for the economic advancement of the country are being converted from virgin to agricultural land and planted under monoculture with an introduced commercial crop. If in such areas centrally lying selected blocks are put aside as reserves, this will serve the interest of both: agriculture and conservation. I will demonstrate this on figures 1 and 2, and quote the relevant paragraphs of my paper verbatim:

"As an example we take a resettlement area where 35,000 acres of virgin land are cleared (to be planted under tea, coffee, cacao or other introduced commercial crop) and this area is divided into 25 equal sized blocks. If such a large area is converted in a short time from virgin land to cultivated country, this can have various effects on the populations of both phytophagous insects and their enemies (parasites and predators). To quote Sir Boris Uvarov's words: "this can lead to a complete reconstruction of the local fauna. Many species are eliminated but a few, favoured by the crops introduced, rapidly increase in numbers. "(Uvarov 1965). Some species which were known as minor pests of the introduced crop in other parts of the country, become major pests. They are not followed by their natural enemies under the changed ecological conditions. The result is that those which were minor pests and turned to major pests become very serious pests to the introduced crop. Various examples of this kind were mentioned by Sir Boris Uvarov at a meeting of the Section of Agricultural Entomology during the Twelfth International Congress of Entomology in London in July 1964 which was also attended by the writer. This mainly happens in developing countries where much virgin land is being opened up for cultivatiom to increase the national income and standard of living of the people." (Szent-Ivany 1967b).

If in the case of the above-mentioned example a phytophagous insect rapidly becomes adapted to the introduced crop and turns to become a very serious pest, some population reduction of the pest may occur in the areas of the peripherial blocks (1-5, 6, 10, 11, 15, 16, 20, 21-25 in Fig. 1.) which border upon the virgin rainforest, as a result of invasion by natural enemies. However, the other parts of these blocks and the central blocks may still have very dense populations of the serious pest causing severe defoliation of the introduced crop. If, however, blocks 7, 9, 17 and 19 are left in their natural conditions and declared total reserves, this will give a chance to the natural enemies of the serious crop pest to build up their populations and they may invade the other blocks. (See Fig. 2.). If the area of 35,000 acres includes habitats of one of our beautiful Ornithoptera species - one of which, O. alexandrae (Rothschild) is the largest butterfly of the world - by declaring blocks 7, 9, 17 and 19 total fauna and flora reserves we serve two interests: that of the agriculturist and that of the conservationist. The settlers will enjoy the beneficial activities of the crop pests' natural enemies and the occasional visit of the magnificent birdwing butterflies to their ornametal gardens.

May I conclude this short address with the statement that the beautiful and rare insects are as much a part of Papua New Guinea's cultural property as the birds of paradise, attractive wood carvings, rock carvings or limestone caves.

It is hoped that in the future, when various areas will be selected as national parks, reserves and sanctuaries, some of these will be devoted to the protection of Papua New Guinea's unique insect fauna

----

R E F E R E N C E S

Szent-Ivany, J.J.H. 1967a. Insect collection and insect protection. In: *Annual Report of the Trustees of the Public Museum and Art Gallery of Papua and New Guinea for the year 1966*: 18 - 24, 2 fig.

Szent-Ivany, J.J.H. 1967b. Some aspects of entomological research with special reference to more recent history in Papua New Guinea. Presidential address, delivered at the Annual General Meeting of the Papua New Guinea Scientific Society on 31st March, 1966. *Rep Proc. Papua New Guinea Scient. Soc.*, 1966 (vol. 18.): 60 - 90.

Uvarov, Sir Boris P. 1964 (1965). The formation of pest fauna by agricultural development in new countries. Proc. XII. Intern. *Cong. Ent.*: 576.

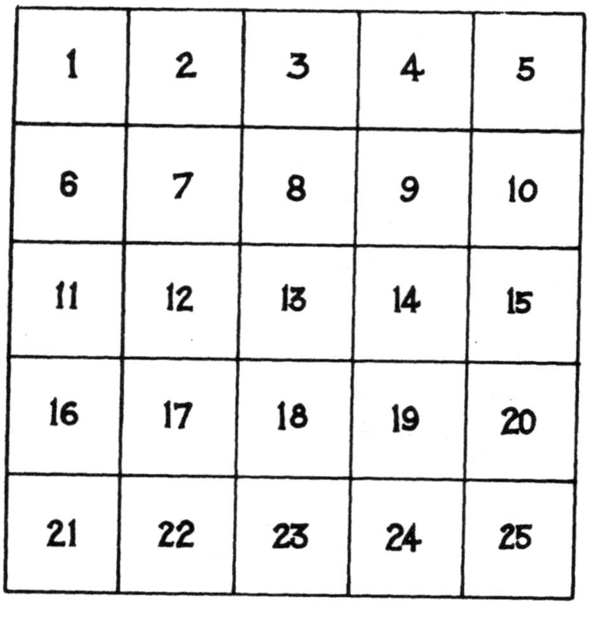

Fig 1

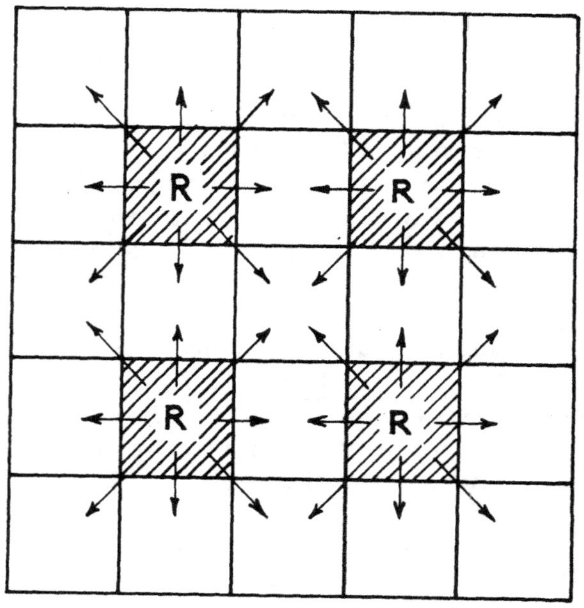

Fig. 2

ECOLOGY AND INSECT PESTS

by

K.P. Lamb
Biology Department
U.P.N.G.

When talking to farmers about their problems (whether in Papua New Guinea, Australia or New Zealand) I have been struck by a recurring criticism of entomologists. They say that entomologists are only interested in finding out a name for the insect attacking their crops and after that they are not particularly interested. I suppose this reflects one important interest of the economic entomologist - and indeed of all entomologists - to put a name to the material being studied. We all know the advantages of this: access to literature, the means of identification for future reference and so on. In a developing country like Papua New Guinea identification is a very real problem as a substantial proportion of our fauna is inadequately studied or quite unstudied. This is perhaps reflected in the fact that of the 436 papers produced in recent years by the Wau Ecology Institute and its associates, 96.3% have been taxonomic, 6% have been biogeographic, 4.6% economic, 1.4% ecological and 0.9% concerned with behaviour. The applied aspects and the ecological aspects of study of our fauna have indeed received little attention. I have not attempted to analyse the papers produced by government scientists in this country; however, one must note that amongst these are some very fine examples by some of our older entomologists. I would single out Dr Szent-Ivany for mention here for his approach to insect pests on a very broad front: ecological and behavioural as well as purely systematic.

Now that Independence has come to this country we face some major challenges in the field of agricultural science and not the least of these challenges is the need for greatly increased food production within the country. I anticipate that one of the major problems associated with increasing our food production will be problems brought about by pests and diseases. It is unfortunate that food crops have been greatly neglected in the past in favour of studies of plantation crops which are of more direct relevance to the export economy. If however we are to feed the people we are going to have to face up to a number of increasingly serious pest problems. Some of these have already come to light in certain areas, for example where experiments have been carried out in rice growing. As reported to the Food Crops Conference in Lae last year, some areas are now considered to be totally unsuitable for rice production because of the extremely serious damage done by some insect pests such as *Leptocorisa*. I would suggest that if we are going to solve these problems we must greatly intensify studies on insect pests, and not merely their systematics but particularly their ecology, the details of their life history and behaviour, their interactions with other species of insects, with their host plants and with various predators and parasites, as well as the physical environment. Some people may think this rather an unexciting prospect but personally I believe it offers a great scientific challenge.

The entomological problems of subsistence agriculture have been greatly neglected. Here we have a situation where we are dealing with a diversity of crops grown by traditional practices within relatively small areas and for limited periods of time. There is, I think, good evidence to suggest that one of the main reasons for shifting agriculture is to avoid the build-up of populations of insect pests and the accumulation of reservoirs of plant disease. Be that as it may, it would seem that subsistence crops in general have many fewer insect problems than medium scale or large scale monoculture.

Apart from the mobility of gardens there are other reasons which almost certainly contribute to this state of affairs. One is that the selection of planting material by indigenous farmers over many centuries has led to a selection of varieties with high tolerance to plant diseases and insect pests or even in some cases to significant degrees of resistance towards pests and diseases. In our enthusiasm to introduce crops of higher yields, improved flavour or improved nutritional value, we must beware of blundering into a situation where we introduce highly susceptible crops which will be quite unsuccessful in the new environment in which they are placed.

Another important area of difference between temperate large scale agriculture and tropical agriculture in developing countries is the degree of reliance that can be placed upon the use of insecticides in these very different situations.

In temperate countries we are accustomed to highly mechanised agriculture and high energy inputs, to the use of large quantities of insecticides applied more or less efficiently with complex machinery at properly scheduled times and at precisely measured doses. Many of these requirements cannot be met in a developing country. There is a very limited availability of machinery and only the simplest of machinery can be maintained in rural situations. The education of many of the farmers is rudimentary and it is exceedingly difficult to ensure that proper safety steps are taken in dealing with hazardous concentrated insecticides or fungicides. Furthermore, the high temperatures and humidities encountered in plantations in the lowland tropics militate against the use of protective clothing so that spray operators tend to be more exposed to spray drift and indeed may be more susceptible to the vapours of toxic materials because of higher temperatures. Thus insecticide application must be limited to substances of very low toxicity to man and which can be used with a wide range of safety of dosage on the plants. Ideally, we would like to get away from the use of toxic insecticides and make their use absolutely minimal wherever possible.

What we need is integrated pest control and the improved total management of agro-ecosystems. If we try to apply this approach we immediately find we have a dearth of even the most basic information on the ecology of tropical crops and their pests and diseases. I would suggest that this is a major field of investigation for the next few years. We should explore to the full the possibility of non-insecticide methods of insect control, whether by cultural or biological control or by the use of resistant and tolerant varieties. As an example of cultural control, let me cite some examples which have come to light through the work of Mr Sands, formerly studying rice at Bubia with the D.A.S.F. The rice bug, _Leptocorisa_, as indicated earlier is a very serious pest of rice which attacks the developing seeds in the milky stage and renders them worthless. Most of the _Leptocorisa_ adults enter the crops from outside and breeding of some species within the crops is a later stage of pest infestation. There are certain species of grasses which are highly favoured hosts for the bugs, in particular _Echinochloa colonum_. The clearing of weeds, including _Echinochloa_ around the edge of plantings of rice will greatly reduce populations moving into the crop. Several other species of grasses were found to be favoured hosts of _Leptocorisa_, though less so than _Echinochloa_. So here is a situation where land usage, land clearing and cultivation can have a significant effect on pest numbers in a crop. On the other hand, high stands of cane grass were found to be repellent or unattractive to _Leptocorisa_ and this plant could very well be used as a border plant to reduce the flight of _Leptocorisa_ adults into the rice crop. Here we are dealing with the interaction of a pest species with several plant species including our crop plant. When we consider the activities of polyphagous insect pests with different tastes for different plant species and plant varieties we find ourselves in even more complex situations. I believe that it would be a very profitable area of study for a group of ecologists, entomologists and others to concentrate on the ecology of insect pests on food crops, with particular reference to subsistence type farming and small to medium size areas of mixed cropping. I believe that there are many new things to be discovered about these agro-ecosystems and new methods of pest control waiting to be explored. We already have considerable evidence that biological control is more effective where several plant species occur together to provide shelter and auxiliary food for the parasites and predators.

Last year I proposed that the time was ripe for the establishment of an internationally funded Institute of Tropical Pest research in this country, with the aim of greatly intensifying field research on food crops with special reference to subsistence agriculture. I believe this to be a project of great relevance not only for Papua New Guinea but also for the neighbouring countries of South-east Asia. While this suggestion was supported by the Food Crop Conference in Lae last year, it was rejected by the Minister of Agriculture on the grounds that U.N. policy was to support single crop oriented research institutes such as the International Rice Research Institute in the Philippines. I must say that I am not in agreement with this policy. While recognising the magnificent work done by I.R.R.I., one must also point to some of the problems that have arisen through failure to place its work into perspective with the subsistence and small scale farmers of South-east Asia. This has on some occasions led to a rejection of rice varieties on account of their economic or cooking characteristics, and it has also made substantial demands on inputs of fertilizers and pesticides which are often beyond the capacity of peasant farmers. I would like to re-emphasise the need for the study of complex agro-ecosystems of the subsistence food crop type. The study of single crop pests belongs to the era of colonial plantation agriculture. Even then it was not a particularl successful policy as evidenced by the persistent economic importance of a handful of major pests which have ravaged important tropical crops for the last century. I believe that an approach to complex food plant agro-ecosystems will need the combined efforts of competent entomologists, ecologists, economists and rural sociologis for maximum efficiency. I do not think that work of this kind is within the capacity of our existing organisations whether government or private but I do think that organisations such as the Wau Ecology Institute may well have a role to play in the study of some aspects of these complex interactions between insects and their host plants. In other words, I would like to see the word ecology underlined in the Wau Ecology Institute in the future.

It is, I think, extremely important to minimise the use of insecticides in agriculture not only to reduce toxic residues in human and animal food but also to reduce costs and to alleviate the substantial toxic hazards to people applying pesticides. There is still a great deal to be learnt about cultural control for almost every pest. If I may quote again from the unpublished work of Mr Sands, his studies on flight behaviour of the sorghum midge have shown clearly that a 10 metre strip cleared of grasses around the edge of a crop will greatly reduce the flight of sorghum midge into the crop. Providing this barrier strip is kept clear or planted with non-grasses (such as peanuts) the adult midges will not fly across the barrier. We need much more work of this kind: the sort of work that has to be carried out in the field where the crops are being grown. I hope that the new race of Papua New Guinean entomologists and ecologists which we are producing will face up to some of these problems and will call upon the expert assistance at present available from overseas workers. Instead of being a scientific backwater, I would like to see Papua New Guinea leading the countries of the Third World in vigorous and forward-looking researches into the problems of food crop entomology and the management of food crop agro-ecosystems

---

The author is indebeted to Mr D.P. Sands for permission to quote some of his unpublished thesis results.

APPENDIX 1 : LIST OF CONTRIBUTORS

Dr. D.H. DALE, P.N.G. University of Technology, Lae.

Dr. A.L. DAHL, South Pacific Commission, Noumea, New Caledonia.

Prof. J. DAVIDSON, P.N.G. University of Technology, Lae.

T.L. FENNER, Dept. of Primary Industry, Konedobu.

Dr. J.L. GRESSITT, Wau Ecology Institute, Wau.

A.K. HAINES, Dept. of Primary Industry, Kanudi.

Dr. L. HILL, University of Papua New Guinea, Port Moresby.

M.A. HILL, National Parks Service, Port Moresby.

Dr. R.J. JOHNS, P.N.G. Forestry College, Bulolo.

Dr. I.M. JOHNSTONE, University of Papua New Guinea, Port Moresby.

Dr. T.N. KHAN, University of Papua New Guinea Port Moresby.

A.A. KIRK, University of Papua New Guinea, Port Moresby.

N. KWAPENA, Dept. of Primary Industry, Port Moresby.

Prof. K.P. LAMB, University of Papua New Guinea Port Moresby. (Symposium Chairman)

Dr. A.P. MACKEY, University of Papua New Guinea, Port Moresby.

J.I. MENZIES, University of Papua New Guinea Port Moresby.

Dr. J.L. MUNRO, University of Papua New Guinea, Port Moresby.

Dr. J.P. SPRADBURY, C.S.I.R.O., Port Moresby.

Dr. J.J.H. SZENT-IVANY, Adelaide, South Australia

Hon. S. TAGO, MP, Minister for Conservation and the Environment, Port Moresby.

K.J. WHITE, Dept. of Forests, Port Moresby.

APPENDIX 2 : BRIEF HISTORY OF THE WAU ECOLOGY INSTITUTE.

Wau Ecology Institute is an outcome of what was originally started as a temporary field station of the Entomology Department of Bishop Museum (of Honolulu), in June 1961. At that time a house (present house No.2 by the labs.) was rented. Two years later it was purchased and at that time a start was made on the arboretum on the 5-acre property. Seeds and young plants from neighbouring forests were planted on the grounds, which consisted of kunai, second growth and coffee. Later (1965) a piping system was developed to bring water from a water-race across Kunai Creek. After that a small zoo was started. The station was a base for field work for various projects from Bishop Museum, carried out in New Guinea Territory, Papua, Dutch New Guinea and the Solomons.

Gradually additional adjacent or nearby properties were purchased. Additions were made in 1965 (present house No. 6); 1967 (rest of "old arboretum" including house No. 4); 1970 ("farm" recently sold, and also "No.5"); 1971 (100-acre plantation, including coffee, second growth, kunai and native forest); 1972 (25 acres of largely kunai, the present "lower arboretum"); 1973 (Kadletz, 10 acres); 1974 ("Reed property" in centre of holdings: proposed new zoo area including squatter village). The net result of these transactions leaves the institute with about 60 hectares of contiguous land. This is roughly divided in thirds between arboretum, coffee and native plus second growth forest.

For most of the 10 years of existence as Bishop Museum New Guinea Field Station, Mr. J. Sedlacek and family were stationed on the grounds or based here between field trips. During that period the "old arboretum" of 6 hectares was developed to the point of including many of the local species of trees and other plants, and a collection of local insects was also developed. The old arboretum has just been named the "J. & M. Sedlacek Arboretum". During this same period, J.L. Gressitt, other Bishop Museum staff members, various collaborators such as Dr. and Mrs. J.J.H. Szent-Ivany, graduate students and others based in field work here, or came to carry out research projects, mostly with financial support from Bishop Museum. Professor R. Traub, later of the University of Maryland, had helped initiate in 1957 a major programme with Bishop Museum on the ectoparasites of mammals and birds covering South East Asia through to the Solomons. In 1960-62 this concentrated on Dutch New Guinea snd since then in Papua New Guinea.

In 1970-71 the Smithsonian Institution started an informal affiliation with the coming of Dr. M. Robinson, Mrs. Robinson and assistant, Miss Y. Lubin, for extended behavioural studies of various insects and other animals. Dr. and Mrs. Robinson came again in 1973-74. Also, researchers came from other institutions to use the station as a base for collecting or for field studies. Gradually this came to serve as a method of helping to support the station, as the U.S. federal research grant funding began to decline rapidly about 1970. It was at that time, together with the planning for self-government and independence of Papua New Guinea, that plans were made to form an independent institute with the hope that it could become permanent and not largely dependent upon grants through Bishop Museum. It was with this in mind that the plantation was purchased, with the hope that by growing coffee, fruit and vegetables, and by providing accommodation and other facilities to visiting scientists, the new organization might become self-supporting. As soon as Wau Ecology Institute was incorporated on 21st October 1971, plans were discussed to make hosting of classes and nature tours a major activity, and also for the enlargement of the zoo, and the development of a small museum also of native animals or plants. Unfortunately, these plans were at first slow in coming to fruit because of further reduction of grant funds and funds available from Bishop Museum. Because of the general financial "squeeze", fewer visiting scientists came to use facilities, and fewer groups came for guiding and instruction.

In 1972 the institute office building was constructed. The situation began to further improve by 1973, and in early 1974 the hostel with 10 sleeping rooms was completed, and we were able to adequately host sizeable classes. Also during 1972-73 the arboretum was greatly expanded and some permanent animal cages were constructed. In 1974-75 the coffee factory and staff housing facilities were improved, and the new laboratory practically completed by end of October 1975. The laboratory was named for Dr. and Mrs. J.J.H. Szent-Ivany and dedicated on 3rd November 1975.

During the period of now over 20 years that J.L. Gressit has been doing field work annually in New Guinea, supplemented particularly for the period from fifteen years ago to five years ago by the work of many others sent by Bishop Museum on museum projects or collaborative projects, much material and data have accumulated. During this same period the museum established two major regular journals and two other serial publications, and much of the space in these has been occupied by articles and monographs resulting from the above field work. A list has been compiled of articles relating to Papua New Guinea and resulting from this work, or at least published by Bishop Museum. The list includes well over 400 references. One of these is the bibliography of New Guinea entomology, co-authored by J.L. Gressitt and J.J.H. Szent-Ivany. Most of those 400 references concern entomology, and especially the ectoparasites of mamals and birds, but some also concern mammals, birds, reptiles, amphibians and other land animals, with an increasing trend now towards ecology.

Plans were made after formation of the institute to stress a programme of education directed toward ecology and conservation. The institute has by now hosted a number of visiting classes. Lectures have been presented, primarily for classes from University of Technology and Lae Technical College. In general classes from University of Papua New Guinea have come with their own instructors. During 1975 an agreement was developed with International Voluntary Services to post three teachers to Wau Ecology Institute, to help in the instruction, but primarily to prepare teaching materials, in the form of handbooks, guides, leaflets, and exhibits, pertinent to the biota of Papua New Guinea. One of these teachers, Mr. Martin Simon, arrived in September 1975, and has given lectures, prepared exhibit material and worked on a pidgin glossary of biological terms.

The publication programme has been initiated with the issuing of the First Biennial Report in 1974, Leaflet No. 1, on tree-planting early in 1975, and Handbook No. 1 on Common Frogs of Papua New Guinea (by J. Menzies of University of Papua New Guinea) in late 1975. The illustrations, and much of the text, for a handbook on the mammals has been completed by A. Ziegler at Bishop Museum. J.L. Gressitt and R.W. Hornabrook are shortly completing a handbook on the beetles. It is hoped that others will follow on lizards, snakes, birds, moths, other insects, etc. in fairly rapid order. A climate and weather report is to come.

During 1974 plans were made to encourage other institutions to join Bishop Museum in helping to support, and to make use of, W.E.T. Early in 1975 the Smithsonian Institution, the University of Papua New Guinea and the P.N.G. University of Technology joined as associate sponsors (K2,000 initial payment plus K500 per year). The Smithsonian is making a larger annual contribution. During this period Bishop Museum has been contributing the services of the director, part of pay for other staff, various other services and has been making nearly all the payments on the debts incurred in purchasing the various properties. Also, the Henderson Foundation has been making annual donations (usually US$2,000 per year) which have gone to buying and shipping equipment and books, some of director's travel, and toward the hostel and new laboratory.

Bishop Museum has been classified as Founding Sponsor and the Henderson Foundation as an associate sponsor. The category of Co-sponsor (K20,000 initial payment plus K2,000 per year) is still vacant. An itemization of the total contributions from and through Bishop Museum has been compiled. This lists the equivalent of K210,650 contributed by Bishop Museum to the development, upkeep and staffing of W.E.I., plus an additional K121,000 equivalent in other funds transmitted through the museum - including equipment and vehicles purchased from grants, and donation money collected by the museum mainly for new buildings and partly for purchase and shipment of equipment or books. This adds up to a total of K340,000 in funds almost entirely from U.S.A. - but partly from Japan and other countries. Donations have also come to W.E.I. directly from Australia, New Zealand, South Pacific Commission, Japan and Europe.

Among the primary needs of the institute now are support from, and use by, more institutions of research, education and development. Additional funds are needed for fellowships, traineeships, for publishing, and for additional equipment and books. Without new support there will be no assurance of funds for the future progress of Wau Ecology Institute.

Aside from financial needs a most pressing need is for government backing to assist in the protection of the important study areas that have been used at and near the institute, and which do and will form the bases of much important work on the fauna and flora of Papua New Guinea. It will be a great loss to science, to education, to Papua New Guinea and to W.E.I. if these critical study areas cannot be protected and perpetuated in their present condition. This will require some initiative, some new stands, and some outspoken words and activity from government. It is urgent that the government give some recognition and status to W.E.I.

APPENDIX 3 : MAJOR CONCLUSIONS OF THE SEMINAR

1. Education at lower levels in relation to the environment is necessary. It is also very necessary to inform decision makers at the village level of ecological principals. The consequences of changing practices should be demonstrated by film and radio environmental programmes. The N.B.C. would be prepared to make such programmes providing appropriate information was provided for them. The Education Department proposes to form a group with such interests for the provision of materials.

2. There is a need to conserve representative habitats. The meeting endorsed the sustained yield management policy of the Department of Wildlife rather than a rigorous policy of legal enforcement.

3. There was a need for closer integration of conservation interests between disciplines and need for a national conservation plan or council in Papua New Guinea. These plans should be co-ordinated internationally and regionally in the South Pacific through the South Pacific Commission.

4. A first step towards conservation could be the registration of areas of national importance.

5. There is a need to prepare a list of workers and resources for environmental investigation within Papua New Guinea, perhaps to be published initially in Science in New Guinea.

6. There should be an annual conference on ecology and conservation sponsored by the Wau Ecology Institute. This should be advertised widely in advance and invitations should be extended overseas, for example to United Kingdom bodies concerned with fauna preservation. Vigorous attempts should be made to get more nationals to attend the next meeting.

www.ingramcontent.com/pod-product-compliance
Lightning Source LLC
Jackson TN
JSHW050946170926
108273JS00022B/35